Peculiar Patents

Peculiar Patents

A Collection of Unusual and Interesting Inventions From the Files of The U.S. Patent Office

by Rick Feinberg

A Citadel Press Book

Published by Carol Publishing Group

A Citadel Press Book
Published by Carol Publishing Group
Citadel Press is a registered trademark of Carol Communications, Inc.
Editorial Offices: 600 Madison Avenue, New York, N.Y. 10022
Sales and Distribution Offices: 120 Enterprise Avenue, Secaucus, N.J. 07094
In Canada: Canadian Manda Group, P.O. Box 920, Station U, Toronto,
    Ontario M8Z 5P9
Queries regarding rights and permissions should be addressed to
Carol Publishing Group, 600 Madison Avenue, New York, N.Y. 10022

Carol Publishing Group books are available at special discounts for bulk purchases, sales promotions, fund-raising, or educational purposes. Special editions can be created to specifications. For details, contact Special Sales Department, Carol Publishing Group, 120 Enterprise Avenue, Secaucus, N.J. 07094

Manufactured in the United States of America

10 9 8 7 6 5 4 3 2 1

Library of Congress Cataloging-in-Publication Data

Feinberg, Rick.
    Peculiar patents : a collection of unusual and interesting inventions from the files of the U.S. Patent Office /
by Rick Feinberg.
        p.   cm.
    ISBN 0-8065-1561-9
    1. Inventions—United States.   2. Patents—United States.
I. Title.
T212.F425   1994
608.773—dc20   94-12695
                                                        CIP

Contents

Beauty and Cosmetics   27

Clever Devices   41

Fine Dining   69

Games, Toys, and Entertainment 87

Health and Hygiene 101

Novelties   141

Pets   153

Safety and Security   *165*

Sports and Exercise   *183*

The Better Mousetrap 219

This book is dedicated to the inventors whose work appears on these pages. I hope it will in some small way further the commercial success of their inventions. I have tried to summarize the motivation behind each patent and its principles of operation as accurately as possible by drawing heavily from the inventors' own words in the original patent disclosures. But even so, errors of interpretation may have occurred. If this has happened, let me apologize now.

Introduction

Each year, the U.S. Patent and Trademark Office issues over 100,000 patents. The great majority are granted to employees of large research corporations, but a surprising number are issued to individual inventors who possess few resources other than their own ingenuity. This book is a collection of actual patents that have been issued mainly to these lone basement inventors. It is a look at some of the unusual solutions dreamed up by a fraternity of solitary thinkers to improve the quality of our lives—and maybe get rich in the process. But first, here are the answers to some common questions that you might have about patents.

What Is a Patent?

According to the Constitution of the United States (Article I, Section 8), Congress has the power to promote the progress of science and technology by granting inventors the exclusive rights to their discoveries. In order to exercise this power, Congress created the Patent and Trademark Office in 1802 to issue patents on behalf of the U.S. Government. A patent is actually "the right to exclude others from making, using, or selling" an invention.

How Long Is an Inventor Protected Under a Patent?

Ordinary, or utility, patents are issued for a term of seventeen years from the date the patent is granted, subject to the payment of maintenance fees. Design patents have a term of fourteen years and require no payment of fees to remain in force.

Can Anyone Utilize the Technology of an Expired Patent?

Yes. Each week about 2,000 patents reach the end of their issue life, or expire due to the failure of the inventors to pay maintenance fees. Smart technology watchers are always on the lookout for some invention or process that enters the public domain in this way. In fact, patent disclosures can sometimes be highly detailed and act as a kind of "instruction manual" for others wanting to utilize the technology.

What Can Be Patented?

A patent can be obtained for any new and useful process, machine, method of manufacturing an object, or composition of matter. Nontrivial improvements to existing devices can also be patented. Taken together, these include practically all man-made objects and ways of making them. The Patent and Trademark Office divides patents into several broad categories: General and Mechanical (covering most machines), Chemical (covering chemical processes, food, and biotechnology), Electrical (covering electronic and optical components and circuits), and Plant Patents (covering new varieties of plants and fruits). Design patents are slightly different. They protect only the appearance of an object, not its structure or utilitarian features. For example, athletic shoe and tire companies frequently obtain design patents protecting the pattern of their treads.

Who May Apply for a Patent?

According to the law, only the inventor may apply for a patent, with certain exceptions. If the inventor is dead, the application may be made by the administrator or executor of the estate. If the inventor is insane (not as unlikely as you might think), the application may be made by a guardian. A coinventor or a person with a proprietary interest in an invention may apply on behalf of a missing inventor, or one who refuses to apply. Inventors may prepare their own applications and file for a patent without the aid of a patent agent or attorney. Unless they are familiar with these matters, or study them in detail, however, they may get into considerable difficulty. Even though a patent may be granted, it may fail to adequately protect the invention.

How Do You File a Patent Application?

A patent application is mailed to the Commissioner of Patents and Trademarks and includes: (1) a written document that comprises a specification of the invention (made up of a description and a number of claims as to its uniqueness) and several forms containing declarations from the inventor, (2) a drawing, and (3) the filing fee. It is not uncommon for some or all of the claims to be rejected by the first action of the examiner. Relatively few applications are allowed as filed. Patents are eventually granted for about two out of three applications filed. Note that if the inventor describes the invention in a printed publication, uses the invention publicly, or places it on sale, a patent application must be made within one year or any right to a patent will be lost.

Do You Need an Actual Working Model To Obtain a Patent?

Usually not. Models are not required in most patent applications since the description of the invention and the drawings must be sufficiently complete in themselves to completely disclose the invention. A working model may be required by the examiner if deemed necessary. This is not done very often, but may be requested in the case of applications for patents covering alleged perpetual motion devices.

How Much Does It Cost To Obtain a Patent?

Independent inventors, small businesses, and nonprofit organizations are considered "small entities" and pay only one-half of the standard fees. The current small entity fees for obtaining and maintaining a patent are:

	Utility Patent	Design Patent
Basic filing fee	$ 355	$145
Issue fee	585	205

	Utility Patent	*Design Patent*
Maintenance fee		
Due at 3.5 years	$ 465	None
Due at 7.5 years	935	None
Due at 11.5 years	1,410	None
Minimum total cost	$3,750	$350

Many other fees also apply for appeals, petitions, reexaminations, corrections, etc.

What Does a Patent Actually Look Like?

The complete text of a relatively short patent appears here. It is summarized in plain language in the section on Sports and Exercise.

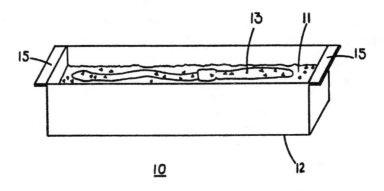

METHOD AND APPARATUS FOR TEMPORARILY IMMOBILIZING AN EARTHWORM

BACKGROUND OF THE INVENTION

1. Technical Field.

This invention relates to the immobilization of live bait for use in fishing. In particular, the invention relates to a method and apparatus for the dewiggling of earthworms.

2. Background Art.

The use of live bait in fishing has long been known to be one of the most effective means for catching fish. The problem with live bait is that any creature has a natural tendency to resist the baiting process. A further complication in the specific case of earthworms is that they are naturally slimy. The ability of the earthworm to curl its body in almost any direction, connected with the fact that it is coated with slimy film, makes it extremely difficult for the fisherman to impale the earthworm with the fishing hook.

GRAHAM. U.S. Pat. No. 2,257,879, discloses a bait box having a compartment that is filled with a dry sand. The worm is dropped into the dry sand which adheres to the worm's body which makes it easier for the fisherman to hold onto the worm. The problem with the method is that the worm is still able to wiggle and curl its body, making it difficult for the fisherman to impale the worm on the fishing hook.

Accordingly, it is the object of this invention to provide a means for immobilizing an earthworm and thereby facilitating the impalement of the earthworm on a fishing hook by the fisherman.

DISCLOSURE OF INVENTION

These objects are accomplished by coating the earthworm with small sharp grained sand. Small sharp grained sand, as opposed to regular dry sand, has a dramatic affect on the worm's ability to curl its body.

A small rectangular container of sufficient length to harbor an earthworm is partially filled with sharp grained sand having a grain size equal or less than 1/20th of an inch. The rectangular container is also fitted with a removable cover which prevents sand spillage during transport. To dewiggle a worm, the fisherman has to simply set the worm in the rectangular container on top of the sharp grained sand. During the worm's natural locomotion process, the sand becomes partially imbedded in the earthworm and causes an immediate reaction wherein the earthworm completely relaxes. The earthworm is then effectively dewiggled and ready to be impaled onto the fishing hook.

Once the sand coated earthworm is immersed in water, the sand rinses free and the earthworm resume its normal wiggly character.

BRIEF DESCRIPTION OF THE DRAWINGS

FIG. 1 is a top perspective view of the container and sand reservoir with a worm.

FIG. 2 is a sectional side view of the container and sand reservoir with a worm.

FIG. 3 is a perspective view of the container cover.

BEST MODE FOR CARRYING OUT INVENTION

Referring to FIGS. 1, 2 and 3, an apparatus for the immobilization of earthworms is generally designated as 10 and is illustrated in its preferred embodiment. The first and only step in the immobilization of an earthworm by the preferred method is to coat the earthworm with small sharp grained sand 11 having a grain size equal to or less than 1/20th of an inch by momentarily depositing earthworm 13 on sand 11.

The preferred apparatus for the immobilization of an earthworm has a reservoir of sharp grained sand 11 having a grain size equal to or less than 1/20th of an inch, and a rectangular container 12 for housing the sand reservoir.

Sand reservoir container 12 is sized for transverse insertion into a standard bait box, not shown. Retainer lips 15 are attached to and extend perpendicularly out from the top edges of the ends of container 12. Retainer lips 15 are sized for cooperative engagement with the top edges of the sides of the bait box, so that when container 12 is transversely inserted into a bait box it is held suspended above the bottom of the bait box which contains a mixture of live worms and humus material.

Cover 14 is contoured to provide for a seal for sand reservoir container 12 and is held in place by the lid of the standard bait box.

To immobilize earthworm 13, one merely deposits earthworm 13 on top of sand 11. During the earthworm's natural locomotion process individual grains of sand 11 become partially imbedded in earthworm 13 and causes an immediate immobilizing reaction in earthworm 13. As a result earthworm 13 will rapidly straighten out and become immobilized. Since earthworm 13 is covered with grains of sand 11, it is not only immobilized, but also easy to pick up and handle.

Once earthworm 13 has been impaled upon the fisherman's hook, not shown, and immersed in water, said 11 washes off earthworm 13 and earthworm 13 will resume wiggling.

While there is shown and described the present preferred embodiment of the invention, it is to be distinctly understood that this invention is not limited thereto but may be variously embodied to practice within the scope of the following claims.

Accordingly, what I claim is:

1. An apparatus for temporarily immobilizing an earthworm which comprises:

a container for housing the a reservoir of sand;

a reservoir of sharp sand having a grain size of 1/20th of an inch or less.

2. The apparatus of claim 1 wherein said container further comprises:

a rectangular shaped container for holding a reservoir of sand, said rectangular container having a length slightly less than the width of a standard bait box;

retainer lips attached to and extending perpendicularly from the ends of said rectangular container for cooperative engagement with the top edges of the sides of a standard bait box for transversely suspending and supporting the rectangular container within the bait box;

a cover for cooperative engagement with the rectangular shaped container for containing the sand.

3. A method for immobilizing an earthworm which comprises partially coating said earthworm with a sharp grained sand having a grain size equal to or less than 1/20th of an inch.

* * * * *

FIG. 1

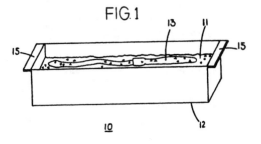

FIG. 2

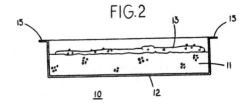

FIG. 3

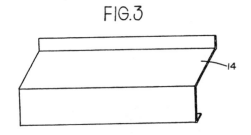

How Do You Obtain a Copy of a Patent?

New and expired patents are announced weekly in the *Official Gazette of the United States Patent and Trademark Office*, which can be found in the public libraries of larger cities. The *Official Gazette* lists one claim and a selected drawing for each patent issued. To view the complete text and artwork of a patent, you can visit one of your state's Patent and Trademark Depository Libraries that maintain patent collections that are open to the public. The scope of these collections varies from library to library, ranging from patents of only recent years to all of the patents issued since 1790. By far the easiest way to obtain a full copy of a patent is to write to:

> Commissioner of Patents and Trademarks
> Washington, D.C. 20231

Include the patent number and a check for $3.00 for each patent you order ($12.00 for each Plant Patent since they are reproduced in color). Be patient. It takes about a month for your order to be processed.

Does Holding a Patent Assure You of Riches?

The latent inventor in each of us secretly believes that holding the patent on something is an instant key to recognition and riches, but unfortunately this is not the case. As most inventors will tell you, inventing is the easy part. Commercializing your idea is much more difficult and time consuming. The Patent and Trademark Office does not assist in the promotion of patents or inventions, other than to permit an inventor to include a notice in the *Official Gazette* that the patent is available for commercial licensing. It has no jurisdiction over questions of infringement and the enforcement of patents, so an inventor must sue patent violators on his or her own. This can be an expensive proposition and may even result in the court's deciding that the patent is not valid!

How Can You Find Out More About Patents?

A good starting place is the publication *General Information Concerning Patents* (ISBN 0-16-041629-9). It contains quite a bit of useful information as well as copies of the forms necessary to file a patent. The book costs $2.25 and can be obtained from your local Government Bookstore or:

> Superintendent of Documents
> U.S. Government Printing Office
> Washington, D.C. 20402

Apparel

Most of us are particular about what we wear, but to inventors, function is usually more important than style. Of course, today's oddity is tomorrow's fashion, so maybe we all can expect to be wearing computerized clothing before too long. The patent literature contains an endless stream of specialized clothing for various occupations and activities. In this section we look at a few examples.

Shark Protector Suit

Patent Number: 4,833,729
Date of Patent: May 30, 1989
Inventors: Nelson C. Fox
Rosetta H. V. G. Fox
both of St. Georges, Bermuda

A shark protector suit is a combined rubber suit and helmet to completely cover the body of the wearer, including a face mask for facial protection. The suit and helmet have a plurality of spikes extending outward to prevent a shark from clamping its jaws over the wearer. Metal plates are attached to the suit at locations where additional protection may be needed. The foam lining of the suit also provides the wearer with additional flotation.

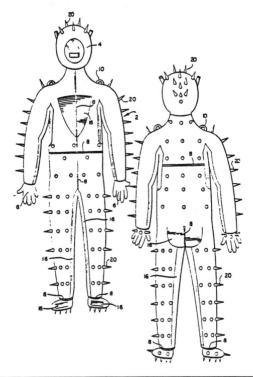

Helium-Filled Sun Shades

Patent Number: 5,076,029
Date of Patent: Dec. 31, 1991
Inventor: Frederick J. Sevilla, Mesa, AZ

A sun shade is formed from two sheets of a tough, flexible material that have been cut to an appropriate shape and size and sealed together at the edges to form a flat, balloonlike structure. Interconnecting channels stitched into the sheets enable the structure to retain its shape without bulging. Helium is filled into the space between the sheets through a valve to make the structure self-supporting without the need for poles or rigid stays. The shade is anchored using cables or other flexible connectors. It is intended to protect outdoors enthusiasts from the dangers of prolonged exposure to the sun during both stationary and mobile activities.

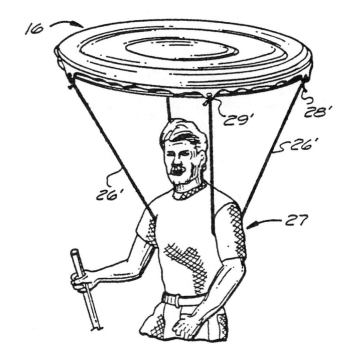

Eyeglass Wipers

Patent Number: 4,789,233
Date of Patent: Dec. 6, 1988
Inventors: Edna M. Arsenault, Ware, MA
George Spector, New York, NY

Eyeglass wipers provide a means for easily maintaining a pair of eyeglasses in a clean condition. Each wiper arm consists of a blade and perforated hollow tube mounted on a pivot connected to the eyeglass frame. The wiper can be made to swing across the surface of the lens by rotating a small knob projecting from the frame. When the knob is depressed, a piston forces cleaning fluid from a reservoir into the wiper arm, which sprays it on the lens. A similar set of wipers can also be positioned to clean the inner surfaces of the lenses.

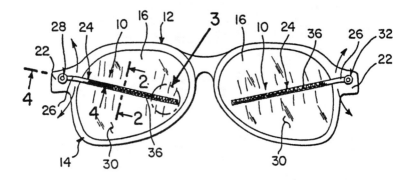

Variable Light–Transmitting Sunglasses

Patent Number: 5,210,552
Date of Patent: May 11, 1993
Inventors: Patrick Baran, Chicago, IL
        Sam Cottone, Wooddale, IL
        George W. Lamping, Barrington, IL

Variable light–transmitting sunglasses allow the wearer to easily change the darkness of the lenses using one finger. The glasses make use of the principle that the amount of light transmitted through two polarizers depends on their relative orientations. A stationary lens of polarized material is mounted in each of the lens frames. A second rotatable polarizing lens is mounted in front of each stationary lens. Gear teeth built into the rotatable lenses allow them to be turned simultaneously when the wearer rotates a gear mounted over the nose bridge. The inventors suggest that in addition to providing easy access for light density adjustment, the exposed gear makes a fashion statement and invites interest in the glasses.

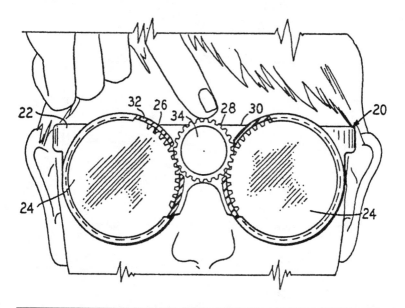

Glasses With Subliminal Message

Patent Number: 5,175,571
Date of Patent: Dec. 29, 1992
Inventors: Faye Tanefsky
Michael R. McCaughey
both of Don Mills, Ontario, Canada

These glasses deliver a subliminal visual message for self-improvement to the wearer. They do this in such a way that although the message is continuously in front of the eyes of the wearer, he or she is not conscious of its presence. The wearer can therefore perform work or engage in sports without interference or distraction. A variety of messages preprinted on transparent adhesive discs can be made available. They are placed in pairs on the inner surfaces of the lenses upward from the normal line of sight (*20*). The stereoscopic properties of human vision cause the two images to merge into one. After ten or twenty seconds, the wearer is no longer conscious of the image and it essentially disappears from sight, but remains present in the subconscious mind.

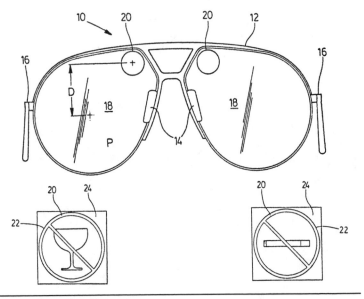

Camouflage Eyeglasses

Patent Number: 4,812,031
Date of Patent: Mar. 14, 1989
Inventor: Tony Evans, Powder Springs, GA

Certain species of game, such as deer and wild turkey, have particularly keen eyesight and are able to detect the whites of a hunter's eyes at considerable distances. Camouflage eyeglasses conceal a hunter's eyes from the prey but do not diminish the hunter's ability to shoot a firearm. Mounted in each eye position is an openweave camouflage netting material having colors and patterns adapted to match and blend into various hunting terrains. The netting mesh is sized to effectively camouflage the eyes of the hunter while not significantly interfering with vision. Unlike conventional sunglasses, no reflections or glints are produced as the hunter moves.

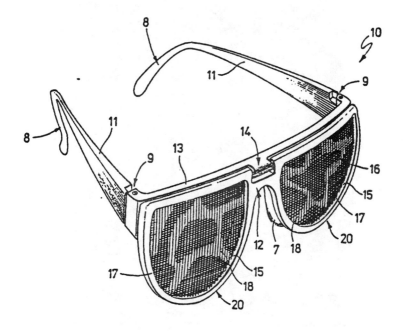

Camouflage Fabric

Patent Number: D. 301,289 (Design Patent)
Date of Patent: May 30, 1989
Inventor: Marybeth A. McIlhinney, Doylestown, PA

There is more to this new camouflage design than meets the eye.

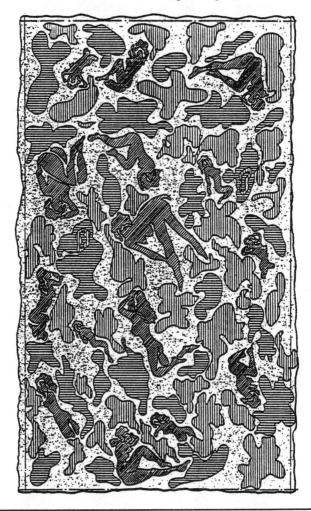

Combined Camouflage and Decoy Device

Patent Number: 5,197,216
Date of Patent: Mar. 30, 1993
Inventor: Raymond E. Norris, Mannford, OK

This combined camouflage and decoy device includes a cap that fits the head of the hunter. Suspended from the cap is a cape that hangs freely and surrounds the entire body of the hunter except for the face. A decoy resembling the bird to be hunted is mounted on the cap. It is constructed from a thin flexible shell filled with a loosely packed lightweight stuffing such as peanut-shaped Styrofoam. Body and head movements of the hunter can impart a realistic motion to the decoy to further increase its effectiveness.

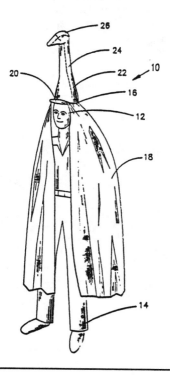

Bulletproof Dress Shirt

Patent Number: 5,008,959
Date of Patent: Apr. 23, 1991
Inventors: Edward A. Coppage, Jr., Oakton, VA
           Richard W. Coppage, Centreville, VA

When worn over a standard dress shirt or dickey with a collar and tie protruding, this bulletproof dress shirt blends into the ensemble and prevents others from discovering the fact that the garment is bulletproof. It features ease of installation, adjustment, and removability through the use of Velcro fasteners. Means to improve comfort by absorption of perspiration are also included. It is of lightweight design and has removable bulletproof pads for ease of laundering. The pads can be made from Kevlar, or a series of layers of an ultrahigh-molecular-weight extended chain polyethylene fabric for superior bullet-stopping power. The shirt may be manufactured with several different front panels of diverse colors and styles so that protection from attack as well as aesthetic satisfaction may be achieved.

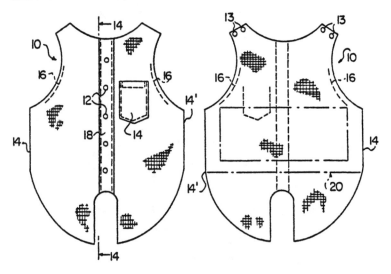

Ear-Mounted Alarm Clock

Patent Number: 4,821,247
Date of Patent: Apr. 11, 1989
Inventor: Reginald M. Grooms, Conway, SC

This personal alarm clock is designed to be worn within or behind the ear for alerting the user at a preset time without detection by others nearby. It allows a person to be awakened from sleep without disturbing a bedmate. The alarm clock housing is constructed from a hard plastic or rubber molded case. The ear tip is formed from a soft material, such as vinyl, silicone plastic, or foam, which is deformable to assume the shape of the ear canal. Due to the close proximity between the alarm mechanism and the user's ear, the alarm sound need not be loud and piercing. A volume control may also be provided.

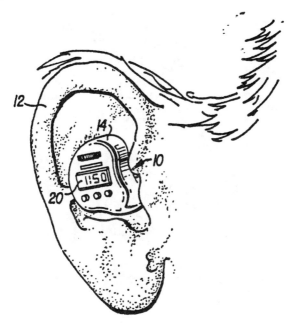

Body-Cooling Apparatus

Patent Number: 4,998,415
Date of Patent: Mar. 12, 1991
Inventor: John D. Larsen, Canoga Park, CA

This body-cooling apparatus is used for removing body or external heat in an environment where sufficient cooling by perspiration is not available. It includes a lightweight battery-powered refrigeration unit (*21*), consisting of a compressor and a condenser. A pressurized liquid coolant such as Freon R114 is fed to a flexible tube network. Cooling takes place by the evaporation of the liquid within the flexible tube network, which is held against the body in a vest and headband or is shaped into a blanket for hospital use. The flow of liquid into the network is controlled by a valve to provide the amount of cooling needed.

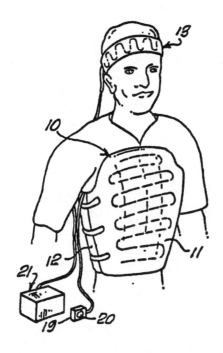

Human Environmental Conditioner

Patent Number: 4,807,447
Date of Patent: Feb. 28, 1989
Inventors: James R. Macdonald, Akron, OH
          George Spector, New York, NY

This environmental conditioning suit is designed to be worn as an undergarment. A turbo pump compressor rotary generator (*18*) worn on the waist circulates refrigerant through multiple conduits in the undergarment to various areas of the body. The compressor requires no external power source since it is actuated by periodic pressure applied to it from the natural breathing of the wearer. A heat exchanger is formed by interweaving the conduits. This keeps the wearer cool in the summer and warm in the winter at a constant body temperature. It therefore allows a person to wear the same type of outer clothing year-round.

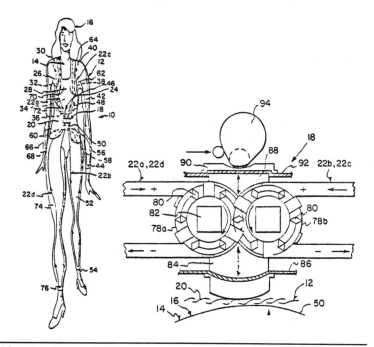

Shoulder Strap Retainer

Patent Number: 4,811,876
Date of Patent: Mar. 14, 1989
Inventor: Michael S. Riggi, Tonawanda, NY

Shoulder bags are a popular accessory, but many people find their use aggravating because of the tendency of shoulder bag straps to slide along and sometimes off the user's shoulder. This shoulder strap retainer provides a base that is attached to the skin of a user's shoulder by means of a suitable adhesive. Different retainers, sized to accommodate shoulder straps of different thicknesses, may be attached to the base. An adhesive would be chosen from the types used for attaching medical bandages and dressings to the skin. This would minimize irritation while ensuring that the retainer remains in place until intentionally removed.

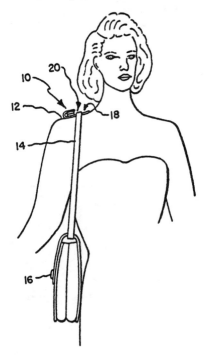

Mixologist Mitt

Patent Number: 4,805,238
Date of Patent: Feb. 21, 1989
Inventor: Cynthia S. Crafts, Pompano Beach, FL

This mitt is designed to prevent a bartender from acquiring a disfiguring callous in the palm of the hand caused by opening a great many bottles of beer in a busy bar. It consists of a flexible, ventilated glove body with a covering to protect the palm. The palm portion is formed into a female toothed receptacle (36) designed to mate with a beverage bottle cap and twist it off.

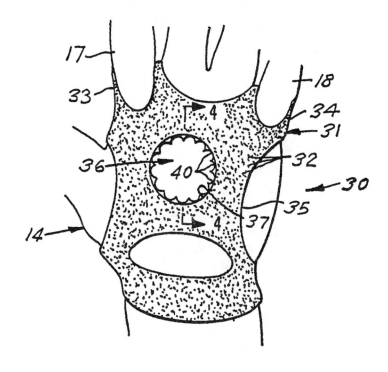

Smoker's Hat

Patent Number: 4,858,627
Date of Patent: Aug. 22, 1989
Inventor: Walter C. Netschert, Buena Park, CA

This portable hat system enables the smoking of tobacco-type products without adversely affecting the environment. Ambient air is drawn into the hat by an integral fan (*64*) and passes upward through the visor (*52*) in front of the smoker's face. Means may be provided to support a lit cigarette under the hat within the intake airflow (*156*). A removable cup (*160*) provides a receptacle for ashes and butts. A filtration, purification, and deionization system (*86, 106*) removes combustion products such as smoke odors and positive ions from the air. An exhaust manifold (*130, 142*) expels the processed air and optionally adds a pleasing scent to it. For the convenience of the smoker, a pocket may be included (*40*) to hold at least one package of cigarettes (*46*).

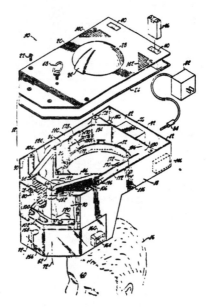

Hatlike Apparatus for Directing Air Flow

Patent Number: 5,085,231
Date of Patent: Feb. 4, 1992
Inventor: Ronald A. Johnson, Union Grove, NC

This hat, like Patent Number 4,858,627 (Smoker's Hat), also contains an air-handling and filtering mechanism for removing tobacco by-products from the discharged air. In addition, its internal fan is reversible, allowing it to provide filtered air, free of pollen and other irritants, to nonsmokers. In this mode of operation, the fan (*32*) draws air over an optional freezable gel for cooling and then through a filter membrane (*16*). The air flows out to the face of the wearer through a conduit (*26*) and a directional outlet (*42*).

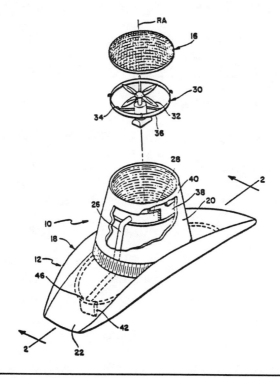

Cap With Hearing-Enhancing Structure

Patent Number: 5,189,265
Date of Patent: Feb. 23, 1993
Inventor: Mark P. Tilkins, Port Washington, WI

This cap includes a pair of curved, tubular hearing-enhancing structures. The cap can be varied in size and the hearing-enhancing structures can be adjusted to accommodate different head sizes and ear positions. The hearing-enhancing structures almost entirely enclose each ear in a conch-shaped cavity, leaving only a forward-facing opening for sound to enter. This concentrates and amplifies sounds originating directly in front of the wearer.

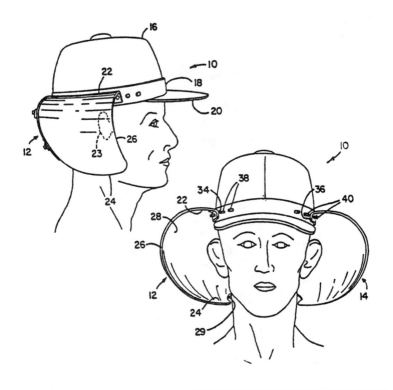

Storage Apparatus for Storing a Necktie

Patent Number: 5,181,670
Date of Patent: Jan. 26, 1993
Inventors: Douglas G. Eaton, Bloomington, IL
         Robert D. Wallace, Bethany, OK

This device allows a necktie to be stored for a substantial length of
time without creasing. In operation, the midpoint of the necktie is
looped over one of the tie rods (*18, 20*) and the whole structure is
inserted into a slot in the transparent cylindrical case (*14*). As the lid
(*16*) is rotated, the necktie becomes wrapped around the tie rods and
is drawn into the case. The entire storage case can be placed in a
suitcase for traveling.

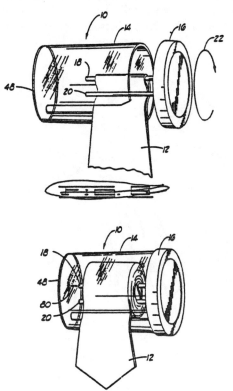

Water-Walking Shoes

Patent Number: 4,801,284
Date of Patent: Jan. 31, 1989
Inventor: Tam T. Lieu, San Jose, CA

A water-walking shoe operates on the principle of a diving bell, trapping air within the bell-shaped shell and utilizing the buoyancy of the trapped air to support the weight of the wearer. It has an outer bell constructed from a rigid plastic material. The foot of the wearer is attached to the outer bell by an adjustable binding (*14, 16*). Vent holes (*22*) are spaced about the periphery of the outer bell. A flexible inner bell (*18*) constructed from a thin plastic sheet material serves to seal the vent holes when weight is put upon the shoe. When the shoe is lifted, the inner bell separates slightly from the outer bell, allowing trapped air to be released. This overcomes the suction effect and permits the wearer to more easily lift the shoe from the surface of the water while walking.

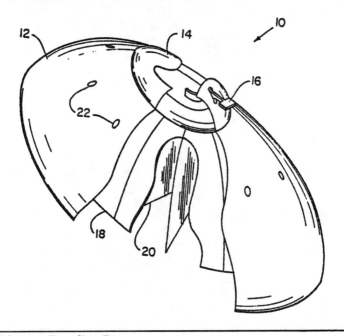

Sandal with Contained Granular Material to Provide a Pad for a Person's Foot

Patent Number: 4,821,431
Date of Patent: Apr. 18, 1989
Inventor: Donald W. Rieffel, Tucson, AZ

This sandal is designed to re-create the original and natural environmental surface that people walked upon before the advent of the universal use of footwear. It allows a person to carry that walking surface with his or her bare feet in urban situations. The sandal contains a channel to receive a person's foot, as well as sand or other granular material. It permits the free movement of the granular material, allowing it to re-form and continually conform to the shape of the wearer's foot in walking. Bristles near the ankle prevent the granular material from being thrown out of the sandal. A powdered desiccant or dry lubricant may be added to the granular material to prevent it from forming clumps or hardening due to moisture present in the sandal or generated by the foot. Deodorants and/or antifungal medication may also be added.

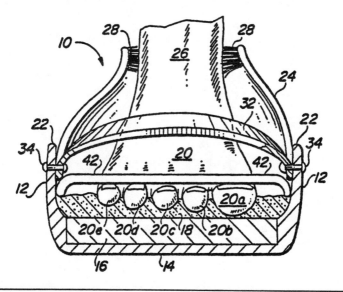

Flashing Footwear

Patent Number: 4,848,009
Date of Patent: July 18, 1989
Inventor: Nicholas A. Rodgers, Perkinsfield, Ontario, Canada

This shoe contains an array of lights around its perimeter for safety or fashion purposes. A mercury switch (*10*) and nine LEDs (*12*) are encapsulated in the running shoe during manufacture. A battery (*16*) is located in a pouch or under a flap inside the shoe. The mercury switch is arranged to be off when the shoe is stationary and horizontal. Motion of the shoe causes the mercury switch to intermittently close and light the LEDs. Additional circuitry can be used to prolong battery life by automatically turning the LEDs off after a preset time if motion ceases but the shoe is not horizontal.

Self-Dusting Insecticide Boot Attachment

Patent Number: 4,881,671
Date of Patent: Nov. 21, 1989
Inventors: Robert D. Horton, Athens, TX
          Doyle J. Farley, Murchison, TX

Insecticide applicators such as flea collars are well-known for use on pets but generally are not used on humans. This self-dusting pouch is secured to a boot to prevent crawling insects such as ticks from passing upward onto the wearer during outdoor activity. The pouch is filled with insecticide through a zippered opening. It is constructed of a porous material and is secured with a Velcro strap to completely encircle the boot.

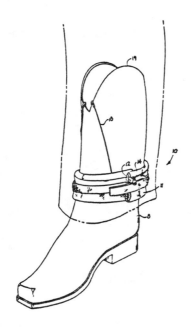

Belt Buckle Having Means for Concealing and Securely Retaining Keys of Different Sizes

Patent Number: 4,905,878
Date of Patent: Mar. 6, 1990
Inventors: William Lovinger
Harry Sielmann
both of Brooklyn, NY

This belt buckle provides a means for concealing and securely retaining keys of different sizes. The key retainer is an elongated hollow housing that maintains the streamlined appearance of the buckle and serves to secure it to the holes in the free end of the belt. Snap buttons on the belt hold the buckle and retain the key in the housing without rattling. When the key is needed, the wearer simply opens the snap buttons and removes the key from the key retainer.

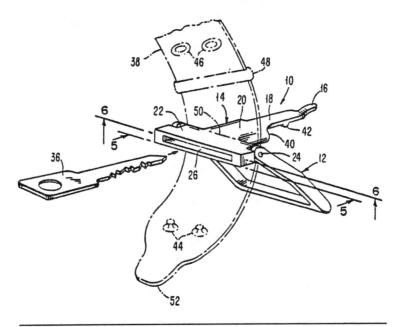

Light-Animated Graphics Display

Patent Number: 4,882,865
Date of Patent: Nov. 28, 1989
Inventor: Frits J. Andeweg, Dallas, TX

This graphics display system may be built into an article of clothing as shown. It consists of a number of light sources, such as Light Emitting Diodes (LEDs) (*17*), that form an integral part of an illustration. The illustration may be printed on a shirt or other surface. A battery-powered (*19*), preprogrammed timing control circuit (*18*) selectively illuminates the LEDs. This produces the appearance of motion of an item in the illustration. In the example shown, the tennis ball would appear to pass back and forth between the penguin players.

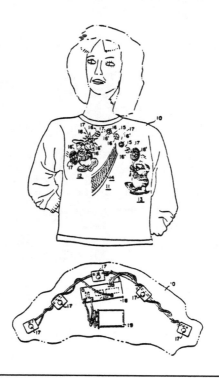

Wooden Cummerbund Apparatus and Methods

Patent Number: 5,232,031
Date of Patent: Aug. 3, 1993
Inventor: Patrick Fredrickson, Marysville, WA

Cummerbunds are usually cloth sashes worn at the waist with formal attire. This novel cummerbund is formed from wood to fit the curvature of the wearer. The wooden cummerbund is cut to shape with a router, from selected stock with an attractive grain pattern. Slots are provided for ventilation. The wood is permanently bent in a mold and steam box or a microwave oven. The distinctive look of the cummerbund may be enhanced with a number of coordinated wood grain accessories such as ties (*30*), cuff links (*40*), belt buckles, and arm bands.

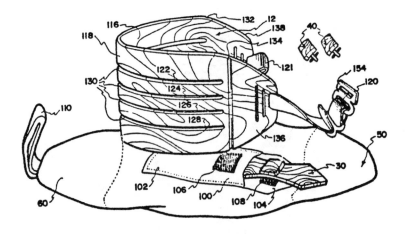

Beauty and Cosmetics

Inventors have not neglected our vanity in their search for new products. Patents have been issued covering just about every possible modification of the human body to make it look more attractive. In this section we find inventions to make body parts look bigger, smaller, hairier, smoother, or just well decorated.

Conductive Cutaneous Coating for Applying Electric Currents for Therapeutic or Beauty Treatment

Patent Number: 5,085,227
Date of Patent: Feb. 4, 1992
Inventor: Gerard Ramon, Gragnague, France

This coating comprises an electrically conductive gel that is applied in a masklike fashion to the face of the user. It is composed of a mixture of polyvinyl alcohol, ethanol, water, and a plastifier that is compatible with the skin. It is used in conjunction with an electrical pulse generator that is connected to two metal plates mounted on opposite ends of an insulated head clip. The conductive plates press against the mask through porous pads soaked with a saline solution. The pads are positioned near the rear of the cheekbones at ear level. The peak voltage of the pulse generator may reach 150 volts at a repetition rate of between 40 and 110 cycles/second.

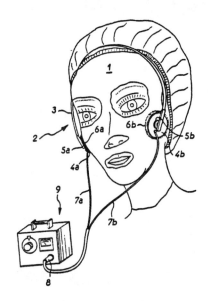

Makeup Method and Device

Patent Number: 4,842,523
Date of Patent: June 27, 1989
Inventors: Jean-Claude Bourdier
Frederic G. Bourdier
Brigitte E. Bourdier
Claude H. Bourdier nee Serre
all of Paris, France

This invention enables a client to use the services of a consulting beautician or makeup artist to provide a personalized style of makeup that the client can reproduce very simply. An instant color slide of the client's face is taken and projected on a removable screen. The slide is reversed to simulate the image seen in a mirror. The beautician applies makeup directly to the screen and also records the type of cosmetics and brushes used on an adjacent chart. The contours of the client's face are sketched onto the screen as well, forming a template for easy reproduction of the makeup style at home.

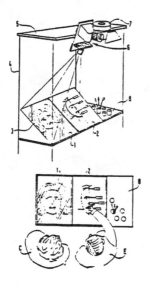

Instant Face Lift

Patent Number: 4,995,379
Date of Patent: Feb. 26, 1991
Inventor: Joan Brooks, New Orleans, LA

This device provides the wearer with a face lift that does not require
surgery. A lightweight plastic mesh band (A) is attached to the scalp
of the wearer with an adhesive. Tabs (3) are attached to the skin of
the face. To make the device as inconspicuous as possible, the band
could be made the same color as the hair and the tabs could be flesh
colored. Straps connecting the band and the tabs (B1, B2) are
tensioned until the desired effect is achieved.

Hydraulic Breast Enhancer

Patent Number: 5,099,830
Date of Patent: Mar. 31, 1992
Inventor: Shunichi Kishimoto, Suita, Japan

This pectoral beauty treatment operates by water pressure to enlarge and/or shape a female breast. It relies on the principle that secretion of the hormone governing breast development can be influenced to a certain extent by actively stimulating the breast. In use, the device is connected to a source of flowing tap water (*26*) that generates a negative pressure within the flow accelerator (*14*) and evacuating chamber (*15*) by means of the well-known Bernoulli Effect. This negative pressure is perceived by the user as a suction within the cup (*3*). When the handle (*22a*) is squeezed, the water stream is directed at the breast, providing a pressing sensation. By alternating the suction and pressing sensations, the desired stimulation can be induced.

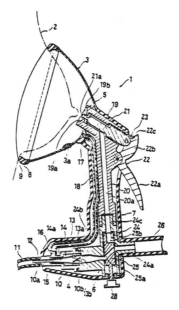

Apparatus to Reduce Wrinkles in Human Breasts

Patent Number: 5,083,555
Date of Patent: Jan. 28, 1992
Inventor: **Ernestine Lewis**, Livermore, CA

This cylindrical bolster pillow is used to reduce or prevent wrinkles in the upper bustline of women who sleep on their sides. It is composed of a resilient foam material and may be covered with a removable fabric casing. It is intended to stay in position as the wearer sleeps on her side, thereby reducing pulling and stretching of the skin of the upper torso.

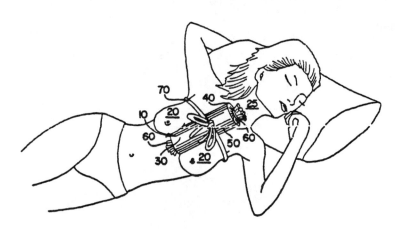

Nipple Ring for Decorating a Human Breast

Patent Number: 5,125,244
Date of Patent: June 30, 1992
Inventor: Hans Zwart, Dayton, OH

According to the inventor, humans have been decorating their nipples from time immemorial by attaching decorative items or jewelry. A common method has been to pierce each nipple and then insert a thread in the pierced hole. Few individuals today, however, are willing to undergo such treatment just to wear jewelry. This easy-to-use compression ring (*10*) may be temporarily expanded by squeezing the end portions (*13, 15*). It secures itself around the base of the nipple by spring pressure. It may be manufactured in a series of different diameters, ranging from 5 to 13 millimeters. The ring can be used to support dangling jewelry or a decorative item that completely covers the nipple, a feature the inventor suggests female wearers might find useful in jurisdictions that prohibit total exposure of their breasts.

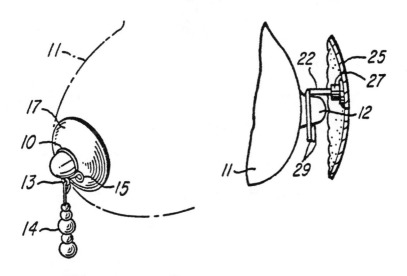

Ear-Flattening Device

Patent Number: 5,076,262
Date of Patent: Dec. 31, 1991
Inventor: Brian M. Coffey, Portland, OR

This ear-flattening device is composed of a pair of flexible pads interconnected by a flexible, one-piece spacer block. The outer side of each pad is coated with a layer of pressure-sensitive adhesive. After a thin protective peel-off layer is removed from each pad, the structure is placed behind the ear of the wearer. The ear is pressed inward toward the head and remains secured in an attractive flattened position. This eliminates the need for cosmetic surgery or the use of an unreliable wad of adhesive tape behind the ear.

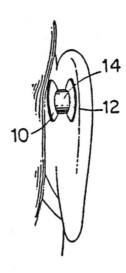

Combined Visor and Hairpiece

Patent Number: D. 302,484 (Design Patent)
Date of Patent: Aug. 1, 1989
Inventor: Dorothy U. Egan, Metairie, LA

The wearer of this combined visor and hairpiece can be protected from the sun and sport a stylish head of hair at the same time.

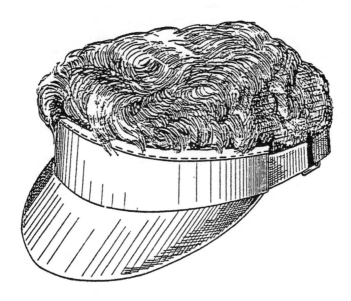

Solar Body Tattoo

Patent Number: 5,052,418
Date of Patent: Oct. 1, 1991
Inventor: David J. Miller, Westmount, Canada

This invention is used to produce a temporary sun-induced skin tattoo. The tattooing device consists of a self-adhesive template with a central cutout corresponding to the desired image. The template is temporarily adhered to the skin and a colored zinc oxide sunblock paste is smeared over it. The template is then carefully peeled off, leaving the desired image "painted" on the skin in sunblock. Tanning rays from the sun (or an ultraviolet lamp) act on the exposed skin and darken it, while the area protected by the sunblock remains light. To ensure a clear tattoo, the user must avoid smudging the sunblock image during the tanning process. The sunblock is simply washed off when the process is complete.

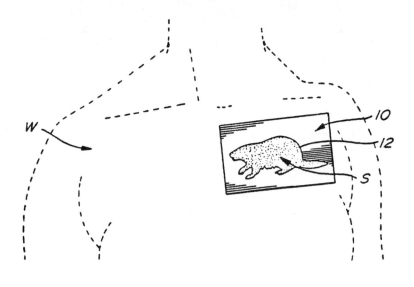

Artificial Fingernail With Clock/Calendar Display

Patent Number: D. 303,161 (Design Patent)
Date of Patent: Aug. 29, 1989
Inventor: Dallas Tomkins, Hawthorne, CA

This artificial fingernail is both attractive and functional. It enables a wearer to look elegant and be on time as well.

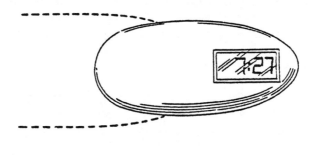

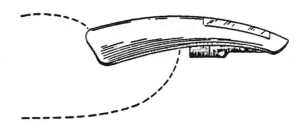

Photographic Imprinting of Artificial Fingernails

Patent Number: 4,974,610
Date of Patent: Dec. 4, 1990
Inventor: Yuko Orsini, Tucson, AZ

In this method of making a set of artificial fingernails, the desired scene image is photographically imprinted on a semirigid layer of photographic film. A layer of transparent plastic film is then laminated over the photographic film. This allows clear viewing while affording additional protection from abrasion or moisture. The individual artificial fingernails are punched out and each is contoured by means of a heated press to better fit the curvature of a natural fingernail. A single scene may span a set of all ten fingernails.

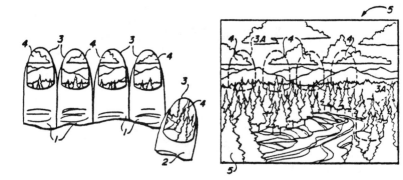

Solar-Controlled Sun Tracker for a Sunbather

Patent Number: 5,211,172
Date of Patent: May 18, 1993
Inventors: **Joseph B. McGuane,** Shirley, MA
Martin J. Lawless, Shirley, MA
Fei-Jain Wu, Chelmsford, MA

This apparatus rotates a sunbather's lounge chair so that the user is optimally exposed to sunlight, even as the sun moves across the sky during the course of the day. It includes two directional photodetectors (*30*) for determining the relative direction of the sun's rays with respect to the chair. Electrical signals from the detectors activate a battery-powered, motorized, rotatable platform (*6*) in the base of the chair. The platform drives the chair clockwise or counterclockwise as necessary to remain centered on the sun. The sunbather therefore obtains an even tan without having to change position. Solar power provides energy to keep the batteries charged. With the addition of microcomputer control, the user could program the exposure desired for both the front and the back of the body, with an audible reminder being given to roll over.

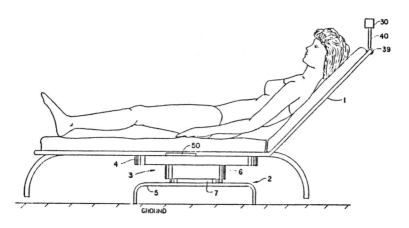

Spray System for Suntanning

Patent Number: 4,846,525
Date of Patent: July 11, 1989
Inventor: Ted A. Manning, Palm Springs, CA

This self-contained spray system is designed to provide a periodic application of suntanning solution for the cooling and comfort of the user. A reservoir consisting of a flexible bladder (20) containing a low-viscosity, water soluble tanning solution is mounted on, or underneath, a lounge chair. Positionable spray nozzles (35) are mounted, using Velcro, at various locations on the rails of the chair. When the user squeezes a rubber bulb pump (30), the spray nozzles discharge a mist of water and tanning solution, which cools the user. Preferably, the nozzles should project a spray upward and inward, resulting in the soft descent of the spray onto the user-occupied area.

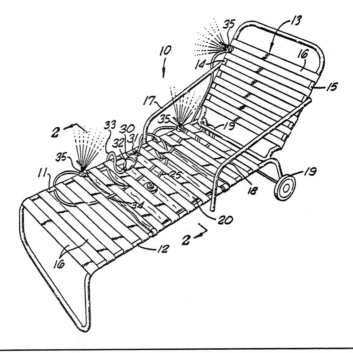

Clever Devices

In this section, we explore some unique solutions to the day-to-day problems of living in the modern world. No facet of daily life is too trivial to escape the notice of inventors trying to help us through our day.

Punishment Wheel

Patent Number: 4,834,657
Date of Patent: May 30, 1989
Inventors: **José R. Gonzalez**, Olympia, WA
         Connie J. Vaughan, Satsop, WA
         Mary Jo R. Gonzalez, Olympia, WA

This apparatus enables a parent to assign a punishment to a child in a fair and objective manner. An assortment of adhesive-backed decals designating various punishments is provided. Decals selected from the assortment are positioned around the base wheel. The child participates in choosing the punishment by spinning a knob and pointer that are mounted in the center of the wheel. When the knob and pointer come to rest, the punishment that is indicated is imposed. From the child's point of view, it appears that an inanimate object is choosing and imposing the punishment, instead of his or her parents. Direct parent-child conflict is thereby eliminated.

Scent Clock Alarm Device

Patent Number: 4,645,353
Date of Patent: Feb. 24, 1987
Inventors: James P. Kavoussi, Brooklyn, NY
　　　　　　Louise D. Hartford, Staten Island, NY

A scent clock alarm awakens the user with a scent instead of a noise or a light. In one form of the invention, a scent disk (*12*) releases a fragrance when either heat or pressure is applied to the surface of the disk by a stylus (*22*). In this way a number of scents may be selected. Another form of the invention utilizes a system of scent cartridges in which the scent source is either solid or liquid, or a combination of both. In both forms, the scent generator ceases abruptly when the alarm is turned off. It is suggested that being awakened in this manner is likely to improve the disposition of the user.

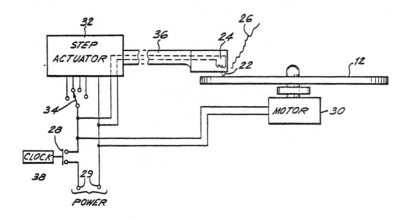

Dripping Sound–Generating Apparatus

Patent Number: 4,852,685

Date of Patent: Aug. 1, 1989

Inventors: **Kengo Maekawa,** Tokoname, Japan
Moitsu, Suzuki, Aichi, Japan
Yukiya Nakano, Aichi, Japan
Hisahiko Tani, Aichi, Japan

This apparatus is capable of providing varied dripping sounds over a long period of time. It simulates the traditional Japanese gardening *suikin-kutsu* (water-harp cave), an arrangement for producing the soothing, echoing sounds of water droplets falling into a pool. A large, inverted ceramic bowl (*1*) enclosing a pool of water forms a sonic resonator. A pump (*7*) delivers water from the pool to a reservoir (*6*), which slowly overflows. The water flow is randomly broken up by a mesh (*10*) as it falls through a hole (*2*) and back into the pool, producing a variety of dripping sounds. The pump adjusting dial (*9*) controls the rate of water flow.

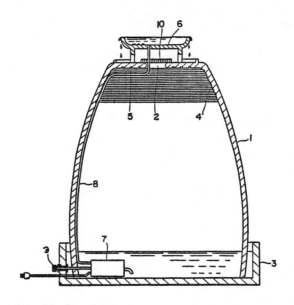

Wrist-Mounted Map Holder Apparatus

Patent Number: 5,183,193
Date of Patent: Feb. 2, 1993
Inventor: Bernie Brandell, Brooklyn, NY

This map holder is mounted to a user's wrist by means of an elastic sleeve (*11*). The sleeve contains a transparent pocket (*12*) that holds a map. Illumination is provided by a bulb (*33*) powered by a battery (in housing *29*) and activated by a switch (*30*). The bulb is attached to a flexible gooseneck support (*32*) for ease of positioning. A similar support (*22*) holds a magnifying lens (*23*) to enable the user to easily read fine details on the map.

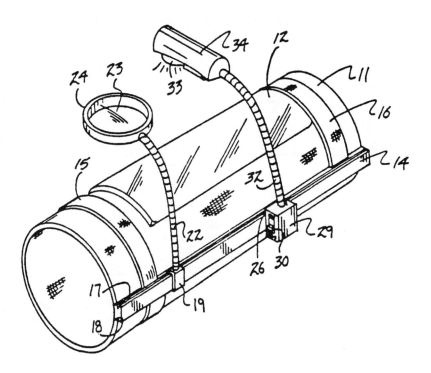

Device for Preventing Snoring

Patent Number: 4,848,360
Date of Patent: July 18, 1989
Inventors: Göte Palsgard   Nils O. Nygren
         Karl-Johan Vikterlöf   Torbjörn Birgning
         Carl-Eric Persson   all of Sweden

This snoring prevention device is entirely contained within a box (4) placed beneath the mattress of the user. A microphone is used to pick up sounds that have passed mainly through the mattress. Electronic filters (6, 8) process the microphone signal in order to determine whether the specific sound frequencies characteristic of snoring are present. Other circuitry (10–14) determines whether the sounds are random, or occur regularly, as is typical for snores. When the snore counter (13) counts a preset number of snores occurring within a given time interval, the sleeper-influencing apparatus (17) is activated. This might be a vibrator or pillow-moving device that causes the sleeper to change position and cease snoring.

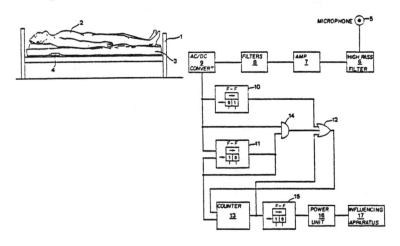

Cushion To Prevent Sleeping on the Abdomen

Patent Number: 5,199,124
Date of Patent: Apr. 6, 1993
Inventor: Daniel E. Klemis, Beverly, MA

This device is intended to prevent a user from sleeping on, or rolling onto, the abdomen during sleep. It consists of a wedge-shaped resilient foam cushion that attaches to the user's chest by means of a belt. The cushion is intended to prevent the aggravation of certain neck and back conditions associated with sleeping on the abdomen. It allows the user to continue to sleep while avoiding an unhealthful position. This is preferable to previous systems, such as securing a tennis ball to the sleeper's chest, which caused the sleeper to wake up from discomfort.

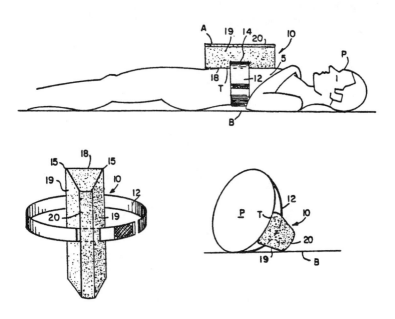

Sound Muffler for Covering the Mouth

Patent Number: 4,834,212
Date of Patent: May 30, 1989
Inventors: Moira J. Figone
           Frank M. Figone
           both of Belmont, CA

According to the inventors, there is a need in our complex society for a device into which a person may yell or scream without disturbing others. This would allow a user to vent built-up anger and frustration in a relatively private manner. The interior of the muffler is constructed of a compliant sound-absorbing foam. Its saddle-shaped opening seals tightly to a user's face. As a further enhancement, a microphone is mounted adjacent to the bottom of the foam saddle in order to pick up a small amount of unabsorbed sound from within the device. The signal from the microphone may be used to activate a light display or meter giving the user immediate visual feedback as to the intensity of sound produced.

Disposable Umbrella

Patent Number: 4,819,679
Date of Patent: Apr. 11, 1989
Inventor: James R. Rex, Pacific Palisades, CA

This disposable umbrella is designed to be strong and easy to assemble in order to afford low-cost protection from extreme glare or inclement weather. It is made of a single sheet of rigid material, such as corrugated cardboard impregnated with a water-resistant polymer starch. The umbrella may be stored as a thin, flat panel and is assembled by folding it into a peaked-roof shape along predetermined creases and engaging a locking tab in the handles.

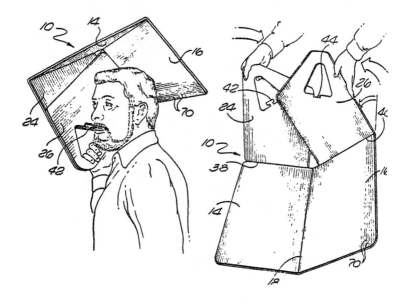

Cap-Shaped Umbrella Held To the Head

Patent Number: 5,201,332
Date of Patent: Apr. 13, 1993
Inventor: Tze-Bing Wu, Taichung City, Taiwan

This cap-shaped umbrella provides protection from the sun or rain while freeing the user's hands to carry other objects. The inventor suggests that this umbrella overcomes the shortcomings of previous designs, such as a tendency to fall from the head, obstruct the line of sight, allow one's pants to get wet, or give the user a headache. He further claims that the mounting bands can be tightly but comfortably fastened to the user's head in the event of strong winds.

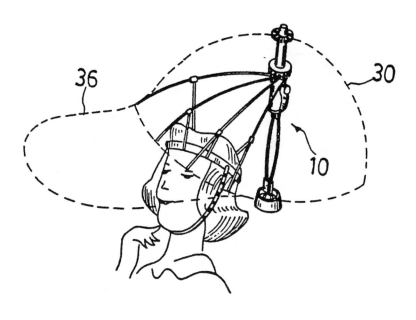

Writing Implement Having Built-in Paper Dispenser

Patent Number: 4,812,069
Date of Patent: Mar. 14, 1989
Inventors: **Kevin R. White,** Cedar Rapids, IA
        Danny R. Gilmore, Denver, CO

This retractable-tip ball-point pen contains a built-in roll of paper that can be dispensed and torn off. The paper roll is mounted in a disposable cartridge (*17*) that lessens the curl of the paper and makes for easy removal and replacement. The cartridge also avoids the necessity for the user to manually load the paper through a narrow slot in the barrel. Rotation of a knob (*18*) connected by gears to a friction roller feeds the free end of the paper (*99*) out through a slot (*29*). A ratchet mechanism prevents the paper end from falling back into the slot after a piece is torn off.

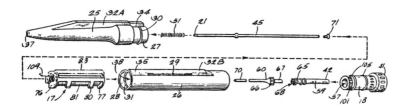

Writing Instrument With Water Level Indicator

Patent Number: 5,190,388
Date of Patent: Mar. 2, 1993
Inventor: Suk H. Lee, Lincolnwood, IL

This device is intended to assist children in easily maintaining a
pencil at the proper inclination when learning to write. It consists of
a transparent liquid-filled vessel that is slipped over the end of a
pencil. With the pencil oriented vertically, a pair of lines marked on
the side of the vessel at an appropriate angle to the horizontal can be
seen. The optimum value of this angle is now believed to be 52 to 55
degrees for best penmanship. When preparing to write, the user tilts
the pencil until the liquid level becomes coincident with one of the
lines. The pencil is then positioned in the user's hand at the proper
inclination for writing. It is effective for either right- or left-handed
writers.

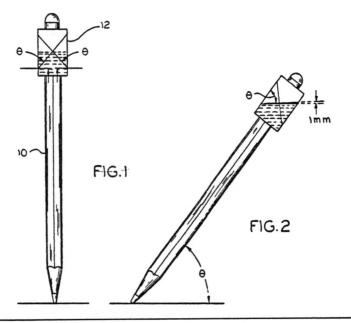

FIG.1

FIG.2

Device for Use by Smokers to Enable Smoking in Public Places

Patent Number: 4,807,646
Date of Patent: Feb. 28, 1989
Inventor: Raphael Sahar, Tiberias, Israel

In this device, a cigarette (*C*) is placed into a short sleeve (*8*) that serves as a holder. A lighter (*16*) drawing fuel from a reservoir (*17*) ignites the cigarette. In use, the smoker sucks on the mouthpiece at the end of the pipe section (*4*). Smoke is drawn through a one-way valve (*15*) and filter (*21*), entering the mouth through an opening (*6*). At the same time, fresh air to support combustion of the cigarette enters through a thin, serpentine tube (*18*). When the smoker exhales, smoke enters another opening (*5*), passes through a one-way valve (*14*), and exits the pipe (at *10*). The smoke remains contained in an expandable bladder (*13*) until it can be vented outdoors.

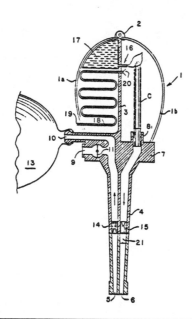

Business Card Dispenser

Patent Number: 4,792,058
Date of Patent: Dec. 20, 1988
Inventor: Robert J. Parker, Martinsburg, WV

This compact dispenser is intended to keep a user's business cards in a crisp, clean condition until needed. It contains an improved mechanism to prevent the accidental discharge of cards within the user's pocket. A single card is ejected when a button (*64*) is slid forward within its slot (*60*). No mention is made as to the maximum number of cards that could be dispensed per minute during a particularly intense networking situation.

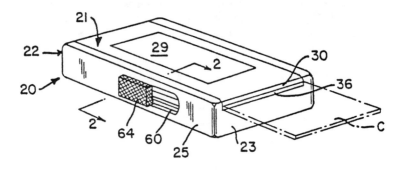

Passenger-Cooling Device

Patent Number: 4,979,430
Date of Patent: Dec. 25, 1990
Inventor: Tallam I. Nguti, Rochester, NY

This device is intended to cool a user within a moving vehicle on a hot and humid day. It may be used within a car, bus, or train, and would be particularly useful when driving within low speed limit areas when insufficient outside air is exchanged with the passenger compartment. The device consists of a plate attached to the forearm of the user and situated in an open window. A breeze may be directed over a selected area of the user's body by manipulating the position and angle of the plate.

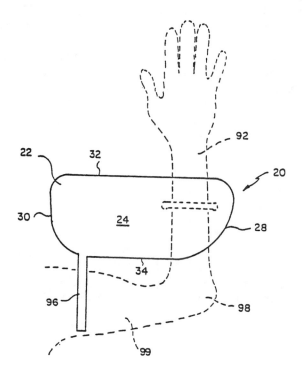

Cremation Apparatus and Method

Patent Number: 4,781,174
Date of Patent: Nov. 1, 1988
Inventor: Kenneth H. Gardner, Bridgetown, Barbados

In this invention, heat rays from the sun are concentrated to a focal point by a reflector (22) in order to cremate a corpse. When the funeral is complete, the coffin (39) is raised on cables (40) or a hydraulic lift (41), leaving the auditorium through a circular hole in the roof. The dramatic effect of the disappearance from view of the coffin is enhanced by the closing of the circular roof light (33), symbolizing death as "the light going out." At roof level, the corpse, with or without coffin, is placed inside a capsule (26). It is then swung over by means of a gantry (25) and lowered to the focal point of the solar collector for cremation. The capsule is also provided with auxiliary burners for expediting the cremation in the event of inclement weather, or due to the pressure of numerous bookings. According to the inventor, this system eliminates the psychologically undesirable connotations of hell fire invoked by sliding a corpse into an oven, or by a vertical descent into the ground. He also claims that it makes possible the realization of man's dream to be reunited with the sun, the ultimate source of all life upon earth.

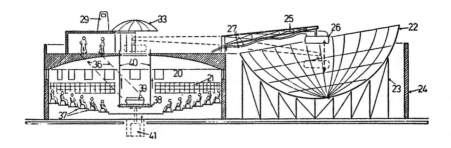

Insect-Capturing Device

Patent Number: 4,817,330
Date of Patent: Apr. 4, 1989
Inventor: Stephen A. Fahringer, Las Vegas, NV

This hand-held portable insect-capturing device requires no batteries and avoids the necessity of touching the insect (alive or dead)
at any time. To prepare the device for use, a bellows (*12*) is squeezed
to expel the air inside of it. The nozzle of the trap (*19*) can then be
positioned next to the insect to be captured. When the trigger is
actuated, a spring (*13*) assists in the rapid expansion of the bellows,
creating a partial vacuum. The insect is drawn into the trap by the
stream of air rushing in to refill the bellows. When the insect comes
into contact with the tacky inner surface of the trap chamber (*20*), it
adheres and thus cannot escape. A conventional insecticide may be
included to exterminate the insect. The trap chamber (*18*) is
disposable and may be thrown away after a number of insects have
been captured.

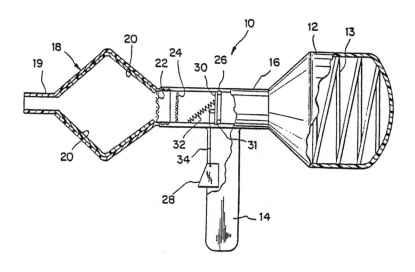

Insect Swatter

Patent Number: 4,787,171
Date of Patent: Nov. 29, 1988
Inventor: Pierre Dagenais, Montreal, Canada

This insect swatter is designed to prevent the soiling of walls and other flat surfaces because it captures an insect without crushing it. The swatter comprises a semiflexible cupshaped body (*24*) with an insect-catching glue-coated sheet lining the inside. The glue-coated sheet can be removed easily and replaced after use. During a strike around the bug, the momentum gained by the semiflexible cup briefly shifts it from a convex to a concave shape. This causes the inner wall to gently touch and adhere to the bug without crushing it. The inventor expects that this system will increase the number of successful strikes. A snap cover can also be provided to muffle the noise produced by a recently caught insect.

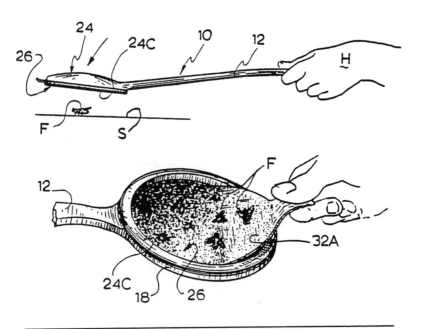

Remote Gas Analyzer for Motor Vehicle Exhaust Emissions Surveillance

Patent Number: 4,924,095
Date of Patent: May 8, 1990
Inventor: Caleb V. Swanson, Jr., Orange, CA

This remote gas analyzer provides a method for quickly and efficiently monitoring the pollutants emitted by a moving vehicle under actual operating conditions. An array of gas absorption analyzer modules (*22*) is stationed on one side of a roadway. Each analyzer emits a beam of light (*10*) that strikes a retro-reflector (*28*) on the opposite side of the road and returns to the module which emitted it. The gas absorption analyzers work on the principle that every pollutant absorbs a unique set of light wavelengths. By determining the wavelengths of the return beam, the specific type and quantity of pollutants in the exhaust can be determined. The array allows a complete sampling of the exhaust plume to be made so that the vehicle's total emissions can be calculated. A camera (*50*) is also included to record the vehicle's license plate number (*48*). Implementing the test system on public roadways, such as freeway onramps, allows motor vehicles to be continuously tested for smog compliance.

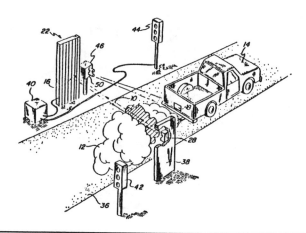

Kit and Method for Demonstrating Courtroom and Trial Procedure

Patent Number: 5,201,660
Date of Patent: Apr. 13, 1993
Inventors: Lynn Copen
Harvey Knapp
both of Kenosha, WI

This portable kit allows a prospective witness to be introduced to the details of courtroom procedure and testimony in a nonthreatening manner. Movable furniture fixtures are provided in order to simulate the exact physical setting of the courtroom to be used. An entire courtroom proceeding may be demonstrated from start to finish by the manipulation of realistic figurines representing the judge, jury, attorneys, witnesses, and other courtroom personnel. The inventors expect it to be particularly useful when preparing children for courtroom testimony.

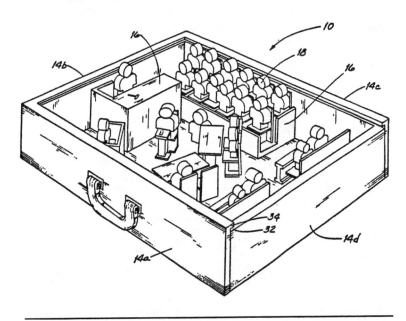

Smoke Hood

Patent Number: 5,214,803
Date of Patent: June 1, 1993
Inventor: David Shichman, Great Neck, NY

This hood is intended to be worn over the head for protection against smoke and gas. The hood is constructed of a flame-resistant transparent plastic that provides face and eye protection without interfering with the visibility or mobility of the user. It is hermetically sealed on all sides and provides an opening into which the head can be introduced. A closure is provided so that the hood may be secured in an airtight manner around the neck of the wearer. The inventor claims that the volume of air contained within the hood will provide a breathable supply of air for five to seven minutes for a normal adult, which can often make the difference between life and death when attempting to escape a burning building. The hood may also be refilled with a fresh supply of air by removing it within a smoke-free area and flexing the bag.

Fetal Speaker System and Support Belt for Maternal Wear

Patent Number: 5,109,421
Date of Patent: Apr. 28, 1992
Inventor: Douglas C. Fox, Van Nuys, CA

This system is composed of a radio or other audio-transmitting device (*21*) mounted on a wide padded belt surrounding the abdomen of the user. A pair of 2.5-inch speakers (*18a, 18b*), mounted within the belt, provide low-volume stereo sound to the developing fetus. The absence of significant external sound emission from the belt allows the mother to wear it without creating a disturbance while socializing or at work. The inventor claims that the objective of the invention is to educate and entertain the fetus, as well as to provide extra support for the mother's abdomen. While supposedly producing a happier, more intelligent, more alert child, it provides the mother with a valuable soothing device to help quiet an upset baby.

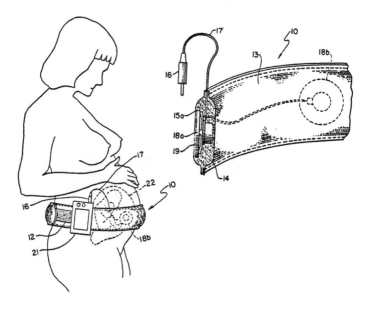

Fetal Communication Device

Patent Number: 4,768,612
Date of Patent: Sep. 6, 1988
Inventor: Janet D. Hodson, Ventura, CA

This device permits spoken communication between a pregnant woman and her fetus. When she speaks into the mouthpiece (*12*), her voice is transmitted through a tube (*14*) to a megaphonelike piece (*16*) that is held flush against her stomach by means of an elastic strap (*70*). It is estimated that a minimum sound intensity of 80 decibels is required outside the abdominal wall in order for the fetus to hear it. The diameter of the tube is chosen to provide resonance to intensify sounds spoken at normal volume (68 decibels) up to this level. This permits the mother to speak in normal conversational tones when using the device. The father may utilize this invention as well, avoiding the necessity of speaking loudly next to the mother's abdomen when he desires to make contact with the fetus.

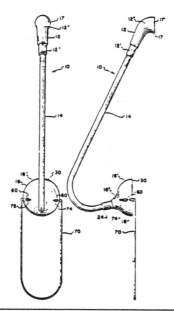

Swimming Pool Escape System for Animals and Insects

Patent Number: 4,972,540
Date of Patent: Nov. 27, 1990
Inventor: James L. Phelps, Ellicott City, MD

This escape system helps prevent the inadvertent drowning of animals and insects in a swimming pool. It also helps the pool owner avoid the distasteful task of removing and disposing of their dead bodies. The system is composed of a weighted base (*20*) placed adjacent to the edge of the pool. A floating ramp is attached to the base by means of a pivot (*60*), allowing it to follow changes in water level. A series of holes and slots in the ramp (*40, 50*) act as a wave suppression system, damping out any swells created by wind or swimming motions of the escapee. This allows the animal or insect to easily climb onto the partially submerged end of the ramp and effect its escape from the water.

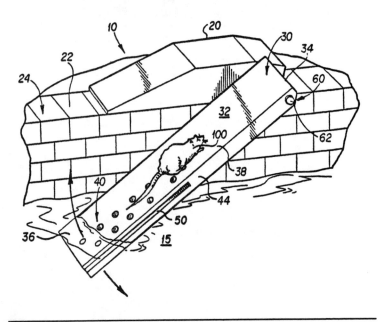

Sun- and Sound-Shielding Arrangement

Patent Number: 5,224,495
Date of Patent: July 6, 1993
Inventor: James H. Robinson, Arvada, CO

This shielding device is intended to allow a user to sleep in a bright
and noisy environment, as might be encountered at the beach. The
sun shield portion comprises an open boxlike shade constructed of a
pliant, flexible material. Additional openings at the corners permit
air to circulate freely around the user's head. Its outer surface is
provided with a reflective coating to minimize heat buildup within
the shade. A strap secures it to the user's head. The earplug portion
comprises a U-shaped springy metal wire covered by a protective
plastic or rubber coating. Cushioned pads on the end of the spring
prevent sound from entering the auditory canal by folding a portion
of the user's outer ear shut. This avoids placing the plugs within the
auditory canal and frees the user from having to clean accumulated
earwax from the device.

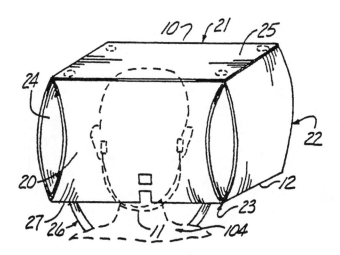

Hands-Free Flashlight Held Between Teeth

Patent Number: 5,226,712
Date of Patent: July 13, 1993
Inventor: Richard G. Lucas, Hollywood, FL

Many individuals who work with their hands in poorly lighted areas (e.g., locksmiths, plumbers, fishermen, etc.) commonly hold a conventional flashlight in their mouths to direct light to a required area. It is difficult to aim a light in this manner or to communicate with fellow workers. Furthermore, holding a flashlight in one's mouth usually interferes with proper swallowing and can result in excessive drooling, which is distracting and messy. This improved flashlight overcomes these problems with an innovative design that includes a rubberized teeth-grasping surface (*60*) and a bite-operated switch (*40*). It is shaped to prevent accidental forced entry into the user's throat. The user can speak and swallow normally while accurately directing light to the work area. A neck strap holds it ready for use.

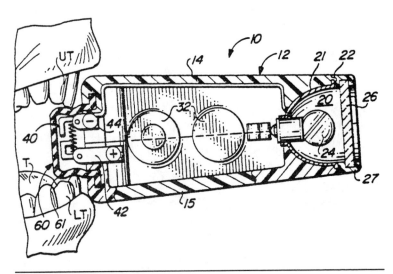

System and Method to Enable Children to Place Their Shoes on the Correct Feet

Patent Number: 5,244,233
Date of Patent: Sep. 14, 1993
Inventor: Mary M. McCraney, Huntington Beach, CA

Young children frequently have difficulty matching their shoes with the correct feet. This color-coded system helps the child overcome a potentially frustrating and uncomfortable situation. Stick-on labels having contrasting colors are placed in each shoe. One or more toenails of the child's right foot are painted the same color as the *R* label placed in the right shoe. The toenails of the left foot are painted the same color as the *L* label placed in the left shoe. It is then a relatively simple task for the child to match the color of the label in a shoe with the color of paint on the toenails of a given foot. Paint may also be applied to fingernails, shoelaces, and socks to provide additional assistance.

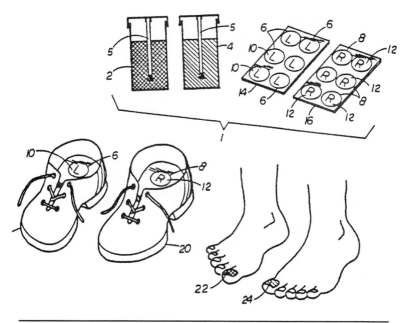

Fine Dining

To inventors, eating is serious business. Where else but in the patent literature would you hear of a nonsettling peanut butter formula referred to as a "Gravitationally Stabilized Peanut Composition"? Patents are regularly issued for new fast foods and eating utensils. Those covering food preparation usually refer to methods for speeding up assembly line operations, but one recent patent disclosed an improved method for carving a turkey! In this section, we find patents to make eating easier, healthier, and more fun. Also, quite a few listings are devoted to helping us eat while on the move.

Crispy Cereal Serving Piece and Method

Patent Number: 4,986,433
Date of Patent: Jan. 22, 1991
Inventor: Alton A. Davis, Dunwoody, GA

This specially designed bowl is used for serving cereal and preventing it from becoming soggy in milk. In use, crispy cereal from the upper bowl is pushed in small portions down the chute and into milk contained in the lower bowl. There, each portion of the cereal can be consumed prior to its becoming soggy. It permits the user to have a leisurely breakfast and still enjoy the benefits of cereal crispness. Furthermore, it can be used with any cereal, not just those that have been treated with a sugar-based crispness-retaining coating.

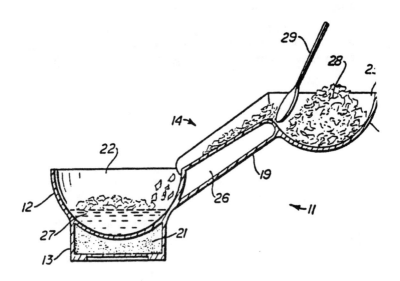

Self-Cooling Beverage Container

Patent Number: 4,669,273
Date of Patent: June 2, 1987
Inventors: Victor H. Fischer, Las Vegas, NV
Dennis A. Thomas, Woodland Hills, CA

A coiled tube insert is provided to cool a beverage within a container. When the container is opened, a pressurized liquid refrigerant is released from the coiled tube into an evaporator. In the evaporator, heat flows from the beverage into the liquid refrigerant, causing it to change into a gas and expand. The heat removed from the beverage causes it to rapidly become chilled within the container.

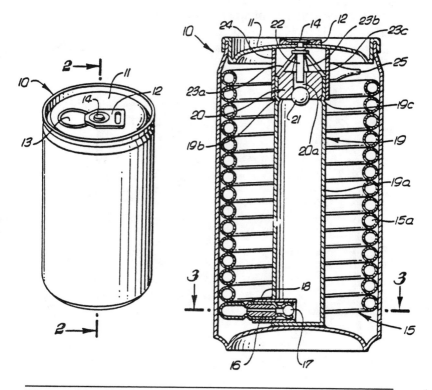

Electrically Warmed Baby Bottle With Rechargeable Battery Recharging System

Patent Number: 5,208,896
Date of Patent: May 4, 1993
Inventor: Alexander Katayev, Richmond Hill, NY

This self-warming baby bottle is constructed from a container made of an electrically nonconductive material such as glass or plastic. Heating wires are embedded within the walls of the container to warm the liquid inside evenly. A rechargeable battery, on-off switch, and temperature-regulating thermostat are mounted in a watertight housing beneath the container. The battery may be recharged without disassembling the watertight housing by mating self-sealing openings in the base with a pronged charging unit. An additional thermometer is provided inside the bottle in order to easily determine the temperature of its contents. The entire unit is dishwasher safe.

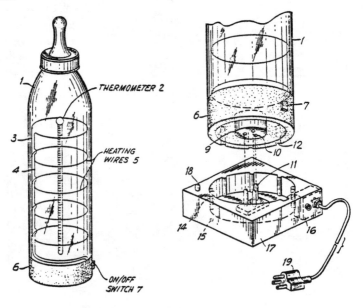

Electrified Tablecloth

Patent Number: 5,107,620
Date of Patent: Apr. 28, 1992
Inventor: Richard E. Mahan, Houston, TX

This electrified tablecloth is designed to prevent crawling insects from gaining access to a user's food or drink. A pair of electrically conductive strips (*18, 19*) with an adhesive backing completely encircle the food-containing area of the cloth. The strips are energized by a 9 volt DC battery (*20*) mounted in a holder (*25*). An insect attempting to traverse the strips will complete the circuit and experience an electrical current through its body sufficient to discourage further travel across the cloth. A person who may accidentally come into contact with the strips usually will not feel the current. At most, a slight tingling sensation may be felt if the person is wet. The electrical apparatus may also be provided in kit form to be installed on a tablecloth of the user's choice.

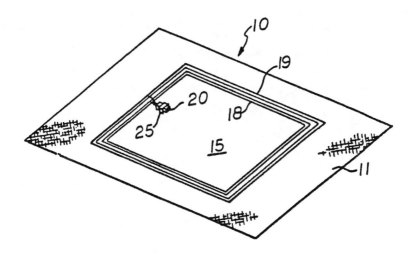

Fly Deterrent Apparatus

Patent Number: 5,003,721
Date of Patent: Apr. 2, 1991
Inventor: James T. Underwood, San Diego, CA

Previous attempts to prevent flies from landing on uncovered food have included devices that create high-speed forced air barriers or that attract the flies toward rapidly rotating extermination propellers. This invention has an elongated flexible wand that rotates at low speed over the food on the tabletop. It is intended to provide a sufficient distraction to a fly to cause it to avoid the area covered by the rotating wand. The wand is slowly turned by a battery-powered motorized base and does not pose a hazard to anyone accidentally struck by it.

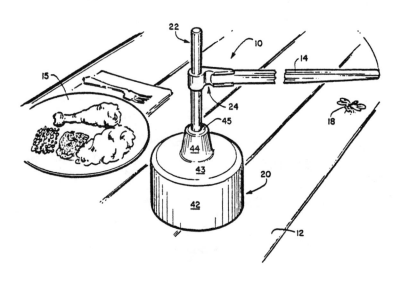

Whole Poultry Egg Analogue Composition and Method

Patent Number: 5,192,566
Date of Patent: Mar. 9, 1993
Inventors: James P. Cox, Jeanne M. Cox, both of Lynden, WA

This invention concerns the creation of an edible egg substitute that has cooking properties identical to those of a natural poultry egg. Unlike some other artificial eggs, it can be hard boiled, soft boiled, scrambled, fried "sunny side up" or "over easy," basted, poached, or used as an ingredient in another food product. The yolk can be engineered to be cholesterol-free and the calorie content reduced to one third that of a natural egg. The yolk is composed of an edible oil or animal fat that has been appropriately colored, flavored, and thickened with a gum or gelatin. After blending, the yolk material is immobilized into a spherical shape by molding and/or freezing. A synthesized protein membrane is formed around the yolk in a setting bath to help it retain its shape. The yolk, including the membrane, may then be placed in and bonded to natural or synthetic egg white, which will keep it from sliding around the skillet while frying. Detailed ingredients and preparation procedures for the egg are given in the text of the patent.

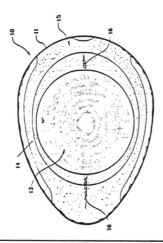

Method and Apparatus for Applying Advertisements to Eggs

Patent Number: 4,843,958
Date of Patent: July 4, 1989
Inventor: Isaac Egosi, Phenix City, AL

This invention relates to a completely new concept in the field of advertising—applying commercial messages to eggshells. Since many people consume eggs on a daily basis, it is suggested that advertisements printed on eggs will be taken into the home and reach a large public with great frequency. The advertisements may be printed in colors and made artistically attractive. Different messages may be printed on eggs within the same carton. The advertisement is applied by positioning the egg on a handling system and while the egg is rotated, applying the advertisement with an ink jet printer (*56b*) to a precise location on the egg for maximum visibility by the customer.

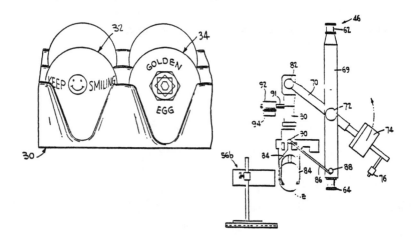

Method for Testing the Freshness of Fish

Patent Number: 4,980,294
Date of Patent: Dec. 25, 1990
Inventors: **Lorne Elias**, Nepean, Canada
           Marek E. Krzymien, Gloucester, Canada

This simple but precise method for determining the freshness of fish relies on a gaseous chemical analysis procedure that can be used instead of the human nose. A sample of the fish to be checked is placed in a closed container at a fixed temperature. Bacterial action in the fish releases a gas known as trimethylamine (TMA), which quickly permeates the container. A sample of air extracted from the container is tested for TMA by a procedure called gas chromatography using a portable instrument. The amount of TMA found is related to the length of time since the fish died. The sampling and analysis are completed in less than five minutes.

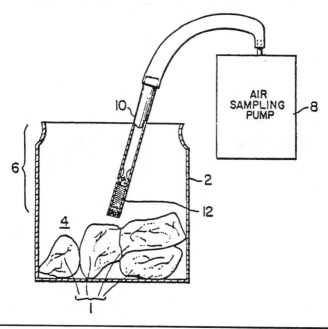

Edible Food Product

Patent Number: 4,919,946
Date of Patent: Apr. 24, 1990
Inventors: Tong S. Pak
Tae S. Pak
both of Statesville, NC

This sandwich consists of two slices of bread or a bun surrounding an edible container. The container may be filled with a large quantity of loose foods, such as noodles, hot dogs, cooked vegetables, fresh salad, mushrooms, barbecue, fruit jam, or a mixture thereof. The container has an outward-turned flange that seals it tightly to the bread, preventing the food from spilling out. This system would also retain large quantities of garnish placed on a hamburger. The edible container itself is made from materials such as potato, corn, wheat flour, and/or rice. It is cooked by broiling, frying in oil, or baking. When made mainly of potato and fried, it provides an improved sandwich that already contains French fried potato chips.

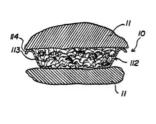
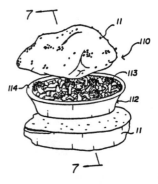

Spaghetti Sipper

Patent Number: 4,942,044
Date of Patent: July 17, 1990
Inventor: Nicholas A. Ruggieri, Rochester, NY

The general public consumes vast quantities of foods while strolling at fairs and shopping malls or just walking down the street. This invention provides a practical means of eating pasta while strolling. The spaghetti sipper is a cup equipped with a removable lid. One end of a piece of cooked pasta is threaded into a tube (*13*) that passes through the lid. The bulk of the pasta is placed in the pasta chamber (*18*). A user draws a serving of pasta into the mouth by exerting suction on the tube. When the desired amount is obtained, the strand is bitten off. A springy pawl mechanism (*14a*) retains the cut end of the spaghetti and prevents it from slipping down the tube into the container. Any of a number of sauces traditionally associated with pasta can be added to the container. Doing so will in fact enhance the operation of the sipper by providing lubrication to the pasta.

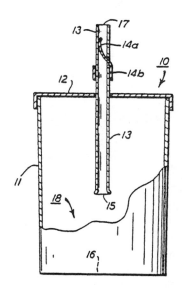

Lobsterlike Food Product and a Process for Producing the Same

Patent Number: 4,889,742
Date of Patent: Dec. 26, 1989
Inventors: Yasuhiko Sasamoto
             Shusaku Hasegawa
             Atsushi Okazaki
             all of Tokyo, Japan

This lobsterlike food product is intended to be wholly edible as well as to have a taste, appearance, and texture quite similar to that of real lobster. It is composed of a fibrous fish paste such as Kamaboko (a fish meat paste product popular in Japan). An imitation muscle consisting of a pair of preprocessed fish products (3) is packed into a concave lobster-shaped mold (10, 11) along with loose ground fish paste. Heat is applied to coagulate the paste into the form of an outer lobster shell (2). A seasoning derived from lobster extract and appropriate food coloring may be included to provide the taste and appearance of a real lobster abdomen.

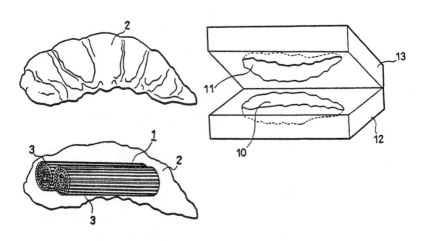

Moo Cream Pitcher

Patent Number: 5,213,234
Date of Patent: May 25, 1993
Inventor: Ioannis Stefanopoulos, Arlington, VA

All currently available pitchers pour liquid from their mouths without any sort of sound effects. This invention, a porcelain, ceramic, or plastic pitcher shaped like a cow, emits a realistic "moo" as it pours cream or milk. When the pitcher is tilted, a gravity-activated switch turns on an electronic device that generates the sound. The sound generator (*3*) is located in the cow's hindquarters.

Flavored Boot for Eyeglasses

Patent Number: 5,202,707
Date of Patent: Apr. 13, 1993
Inventor: Adam S. Halbridge, Houston, TX

Many people who wear eyeglasses frequently remove them and chew or suck upon the temple arms. This activity can leave unsightly indentations and tooth marks on the frames and does not taste particularly good. The inventor suggests that many eyeglass wearers who do chew on the frames of their eyeglasses would enjoy having a desirable flavor imparted to them. This eyeglass boot is constructed of a tubular nontoxic plastic material such as PVC, polyethylene, Teflon, or rubber that has been impregnated with a flavor. Children might have a preference for a fruit or candy flavor, whereas an adult wearer might prefer a spice flavor such as cinnamon or jalapeño. Each flavored boot (*30*) slips snugly over the end portion of a temple arm (*32*) and protects it from damage. In an alternate form of the invention, the boot may be constructed from an edible material such as licorice or taffy that dissolves over time as the wearer chews or sucks it.

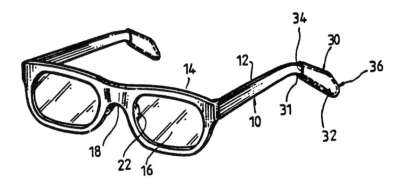

Elastomeric Bladder for Dispensing Ice Cream

Patent Number: 4,798,313
Date of Patent: Jan. 17, 1989
Inventor: Brent L. Farley, Baltimore, MD

This tubular, expandable, elastic container is designed to dispense a frozen confection as it thaws. High-density polypropylene or polyethylene, about one-tenth of a millimeter in thickness, may be used for construction of the container, which is initially molded into a spiral shape. The container is filled by forcing frozen ice cream inside, causing it to temporarily straighten out. As the ice cream thaws, the container slowly curls back into its original shape, creating sufficient pressure to force the thawed ice cream out through an opening (*324*) into the user's mouth. As an additional amusement, by removing one's finger (*362*) and blowing into the mouthpiece, a charge of ice cream in the T-shaped handle (*366*) may be expelled through a hole (*364*) as a projectile. In another version of the invention, finger coverable openings can also be provided so that the device, minus its contents, can be played as a flute or a whistle.

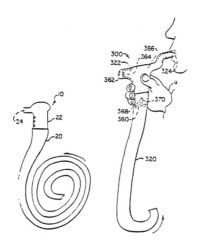

Dripless Ice Cream Holder

Patent Number: 5,224,646
Date of Patent: July 6, 1993
Inventor: Anthony J. Biancosino, Princeton, NJ

This dripless ice cream holder solves the problem of soiling one's clothes or hands while eating an ice cream cone. Ice cream or another frozen confection is placed into a holder in the top portion of a receptacle. The holder may be sized to support a scoop of ice cream or an entire cone. The ice cream that melts before it can be eaten drips down through a number of holes into a catch basin at the bottom. When the main body of ice cream has been consumed, the user tilts the receptacle and drinks the accumulated liquid through a slot. The usual messiness associated with eating an ice cream cone is eliminated and sanitary consumption is achieved because the consumer's hands never touch the food.

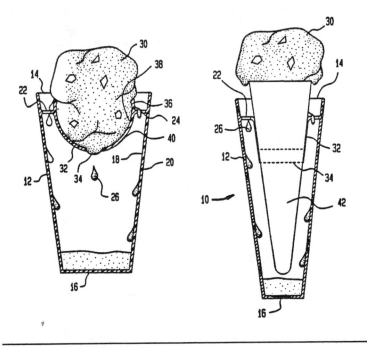

Confection Package

Patent Number: D. 301,835 (Design Patent)
Date of Patent: June 27, 1989
Inventors: Jeffrey B. Gorman
          Marilyn Katz
          both of Los Angeles, CA

This novel package enables a user to eat a confection while satisfying the childhood urge to suck one's thumb.

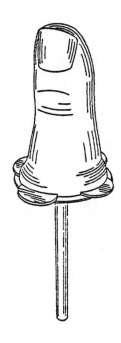

Tool for Imprinting Indicia on Hot Dogs

Patent Number: D. 337,031 (Design Patent)
Date of Patent: July 6, 1993
Inventor: Joseph A. Schroth, Jr., Atlanta, GA

This invention appears to be a sort of "branding iron" for imprinting a message on a hot dog.

Games, Toys, and Entertainment

You can take the measure of a society by how it spends its leisure time. For the most part, we want our toys to be educational, exciting, or just plain fun. Board games have been patented to teach everything from farming to crossing the street, and sometimes the "rule book" takes up many pages of the patent disclosure. In this section, we look at some inventive ways to pass the time.

Amusement Device for a Toilet Bowl or Urinal

Patent Number: 4,773,863
Date of Patent: Sep. 27, 1988
Inventor: Louis R. Douglas, III, San Francisco, CA

In this device, a number of urine-activated pressure and temperature sensors are embedded in a plastic base placed near the drain of a urinal or toilet bowl. Next to each sensor is an associated LED or buzzer that is activated by the sensor. The device may be connected to a video screen or a speaker located above the urinal for providing additional audible or visual stimulation to the user. It is suggested by the inventor that use of this device will help maintain the cleanliness of nightclub rest rooms by interactively engaging the attention of inebriated patrons and helping them to remain on target.

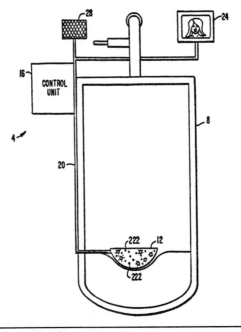

Toy Creature Having a Tongue for Capturing Prey

Patent Number: 4,778,433
Date of Patent: Oct. 18, 1988
Inventors: **Robert S. McKay,** Morton Grove, IL
          Dennis R. Dahm, Streamwood, IL
          Robert S. McLennan, Hillside, IL

This toy simulates the capturing feats of natural predators such as frogs and lizards that use an extended tongue to ensnare their prey. The tongue is constructed from a number of linked sections and normally resides in a coiled position entirely within the closed mouth of the toy creature. When the activating handle (78) is pushed forward, the mouth opens and the tongue uncoils to a substantial length. For added realism, the end of the tongue may be swept around in pursuit of the prey by rotating the handle. When the prey is encountered, the handle is withdrawn, causing the tongue to curl around the prey and draw it into the closing mouth.

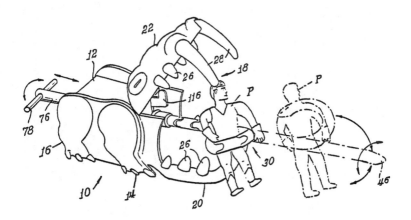

Game for Foretelling Particulars of a Person

Patent Number: 4,779,871
Date of Patent: Oct. 25, 1988
Inventor: Claudia M. Rasmussen, San Diego, CA

There is often speculation at baby showers about the characteristics, attributes, and future particulars of the unborn child. Guests frequently enjoy predicting things such as the gender, date of birth, birth weight, talents, and interests the baby will have. In one version of the game, a poster of a surprised-looking unclothed baby with no specific gender is hung on a wall. Participants select a male or female gender symbol according to their estimation of the unborn baby's sex and write their prediction of the birth date on the back. While blindfolded, each player attempts to place the gender symbol between the legs of the baby in the poster. The unborn baby's gender and birth date are predicted to correspond to the data on the gender symbol that lands closest to the corresponding genital area of the baby in the poster.

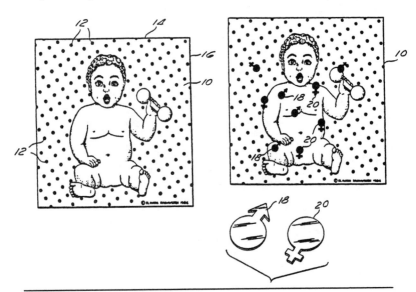

Electronic Game With Animated Host

Patent Number: 4,799,678
Date of Patent: Jan. 24, 1989
Inventors: Rouben T. Terzian, Chicago, IL
  Jeffrey D. Breslow, Highland Park, IL
  Donald A. Rosenwinkel, Oak Park, IL
  John V. Zaruba, Chicago, IL

This electronic game includes an animated game show host character. A base for the character simulates a stage and houses the electronic components. The base also contains a gridwork of indicator lights plus a touch switch pad of thirty-two squares (*50*) and four individual player buttons (*61–64*). Players identify themselves by pressing their initials or the words *Mom* or *Dad* on the keypad. Animation of different features of the character, such as its eyes, head, and arms is powered by a single motor. Another motor moves the lower jaw of the character's mouth in synchronism with a synthesized voice or audio tape. The animated host issues instructions to the players and keeps score during various word- and phrase-guessing games.

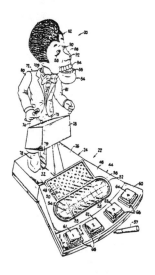

Doll With Simulated Hair Growth

Patent Number: 4,801,286
Date of Patent: Jan. 31, 1989
Inventors: Henry Orenstein, West Caldwell, NJ
　　　　　Joseph J. Wetherell, New York, NY
　　　　　Allan H. Buckwalter, Philadelphia, PA

It generally has been found that toy dolls simulating the actions of human beings have a high level of appeal. This doll contains a movable lock of hair (*16*) that simulates hair growth. With the right arm of the doll in a downward position, the left arm may be wound in a counterclockwise direction. This operates a gear mechanism that winds a cord (*52*) attached to the movable lock of hair onto a spool and retracts the lock into a tubular sleeve (*38*) within the doll's head. At the same time, a biasing spring (*40*) is compressed. Raising the doll's right arm releases a brake, allowing the spring to advance the hair outward from the head at a controlled rate.

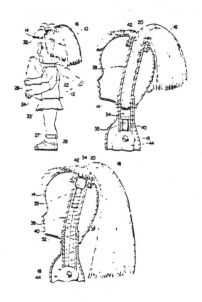

Projectile and Target Game Apparatus

Patent Number: 5,082,291
Date of Patent: Jan. 21, 1992
Inventors: **Mel Appel**, Livingston, NJ
          Denni Rivette, Akron, OH

The projectile ball in this target game is made of a soft, lightweight, resilient material. Its outer surface is covered by a feltlike substance and is unlikely to do damage to a player's skin, regardless of the force with which it is thrown. It is optimally at least 3 inches in diameter so that it cannot accidentally enter the eyes, mouth, ears, or nostrils of a player. A cap covers a major portion of the head of a player and has an outer surface with one or more target areas. These regions are covered by a number of small hooklike projections such as are found in Velcro. The projections can interlock with the felt surface of the ball and temporarily retain it in place. Each target area may be assigned a different score value for the playing of throw-and-catch or throw-and-avoid games.

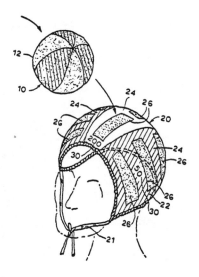

Bicycle-Mounted Water Toy

Patent Number: 4,807,813
Date of Patent: Feb. 28, 1989
Inventor: Larry Coleman, Flint, MI

This toy combines a child's natural love of water pistols and squirt toys with the enjoyment of riding a bicycle. A fluid reservoir (*38*) constructed from one or more tanks is mounted on the handlebars of a bicycle. A pump (*40*) is attached by a mounting bracket to the front fork of the bicycle. When a handlebar-mounted control lever is engaged, a friction wheel projecting from the pump is brought into rolling contact with the front wheel of the bicycle. The wheel is connected to an impeller inside the pump and generates the water pressure necessary to operate the squirt gun. Water is then conveyed through flexible plastic tubing (*46*) to an outlet nozzle (*48*) on top of a padded, impact-resistant helmet worn by the user. The user can direct the stream of water in any direction simply by pointing his or her head to the right or the left. An additional benefit of the toy is that it encourages young bicycle riders to wear safety helmets.

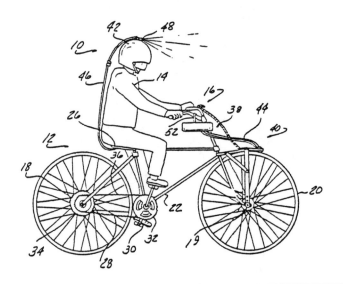

Parlor Game

Patent Number: 4,813,683
Date of Patent: Mar. 21, 1989
Inventors: **Elizabeth F. Ginovsky,** Brockport, NY
      Colleen J. Richenberg, Rochester, NY

This parlor game, called "FANTASM," is played by placing a number of lightweight plastic pucks on the top faces of the blades of a stationary ceiling fan. Inserts attached to the pucks are labeled with point values, double score, bonus points, and free spin indications. When the fan is energized, centrifugal force ejects the pucks in all directions. Players on opposite sides of a center line under the fan attempt to catch as many pucks as possible before they hit the ground. Pucks that have bounced off the walls, floor, furniture, or people are out of play. Players may not grab pucks from the moving fan. The game may be divided into periods consisting of a predetermined number of spins of the fan. Many other variations are possible. In "Reverse FANTASM," Velcro strips are attached to the fan blades and the object is to toss lightweight Velcro covered balls onto the blades. In "Ultimate FANTASM," the game is played in an unlit room with glow-in-the-dark pucks. In "Intimate FANTASM" (for an adult couple), the puck inserts are imprinted with articles of apparel, suggestive of actions to be taken by one or the other player when a puck is caught.

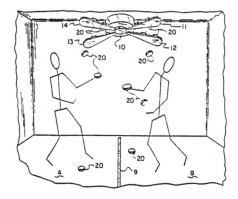

Anatomical Educational Amusement Ride

Patent Number: 4,865,550
Date of Patent: Sep. 12, 1989
Inventor: Shao-Chun Chu, Monterey Park, CA

This amusement ride is housed in a large building having an external appearance simulating a man and a woman resting partially under a blanket. It is intended to educate riders about the human body and how to avoid its various pathological conditions. Riders are taken through a succession of cavities that realistically simulate normal and diseased internal organs of the man and woman. Entrance to the ride is via a stairway (*44*) leading to the mouth of the man (*50*). Riders board a train that passes into the cranial cavity of the woman (*92*), past a display of brain, eye, and ear organs, and into a body portion that is representative of the abdomen of both the man and the woman. From there, it enters a simulated esophagus (*130*), stomach (*134*), and intestines. It goes through the urinary and reproductive tracts, then through a liver (*210*) and a cardiovascular canal (*220*) before exiting through a lung (*270*) and windpipe (*282*).

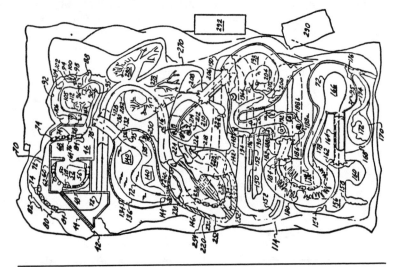

High-Performance Motorcycle Simulator

Patent Number: 4,978,300
Date of Patent: Dec. 18, 1990
Inventors: **Howard Letovsky**, Los Angeles, CA
         Bernard Fried, Beverly Hills, CA

This invention attempts to create a realistic simulation of the sights, sounds, and forces experienced while operating a high-performance motorcycle. The simulator base allows a combination of six independent types of movement: The frame can be swayed side-to-side while simultaneously being surged forward or backward, rotated about a vertical axis (yawed), tipped forward and back (pitched), or inclined from vertical (rolled). The motorcycle-riding experience is enhanced by blowing air past the rider (22) at a wind velocity corresponding to the simulated speed. A computer-controlled video projector (18) displays a simulated view of the riding environment on a rear projection screen (20). A sound system within the rider's helmet accurately simulates road and engine noise. In the event of trouble or a simulated accident, the operator is provided with a failsafe system to shut down the simulation at any time and return the frame to a neutral position.

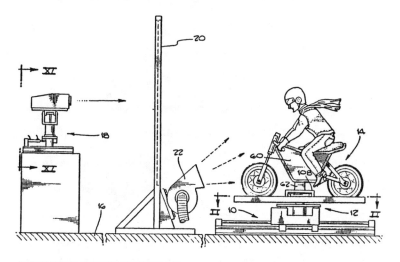

Stuffed Toy With Heater and Phase-Changing Heat Storage

Patent Number: 4,979,923
Date of Patent: Dec. 25, 1990
Inventor: Toshio Tanaka, Yokohama, Japan

This stuffed animal has a built-in heat storage reservoir to maintain it at a relatively constant temperature for long periods of time. The heat storage unit contains a synthetic resin film sack impregnated with about 200 grams of a substance having a high heat storage capacity (*9*). To charge the unit, an electrical cord is connected to a receptacle (*7*) in the toy. This energizes a heater (*5*) in contact with the heat storage material. After about thirty minutes, a thermostat (*11*) causes a buzzer (*12*) to sound, signaling the completion of heating. At this time, the storage material is completely melted and the temperature of the surface of the toy is about 136°. The electrical cord can then be removed and the toy safely given to a child. Heat is given off as the storage material slowly solidifies and the surface of the toy remains at 136–104° for at least five hours.

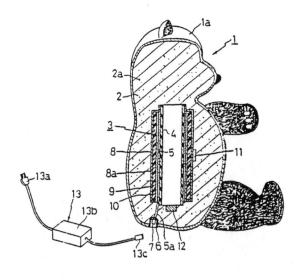

Brain Wave–Directed Amusement Device

Patent Number: 5,213,338
Date of Patent: May 25, 1993
Inventor: Gregory R. Brotz, Sheboygan, WI

This game allows a signal derived from a player's brain wave intensity to direct the movement of a rotating circular display. It can be used for relaxation or amusement, or as a competitive game with other players. The changing visual patterns of the display provide feedback that directly influences brain wave production. Music, or other tones, provided through the headphones can also be altered in response to changes in brain wave activity. When two players compete, the direction of rotation of the display is determined by the player with the stronger brain waves. Each player can also be provided with a "zapper" button to occasionally turn the display blank and disrupt the brain wave production of an opponent.

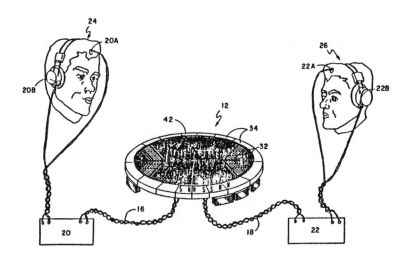

Lover's Game and Method of Play

Patent Number: 5,213,509
Date of Patent: May 25, 1993
Inventor: John C. Gunn, Gold Beach, OR

This game for lovers is intended to be played continuously, forming a part of everyday life. It consists of two containers and a number of small game pieces. The total number of pieces is more than enough to fill one container, but insufficient to fill both containers. The game is played by initially distributing an equal number of game pieces into each player's container. When a player performs a loving act, he or she gives the other player a game piece for storage in the recipient's container. The acts may be sexual favors, kisses, remembering an anniversary or birthday, dinner out, helping around the house, a special present, breakfast in bed, etc. Sooner or later, the pieces may overflow one of the containers, reminding the owner of the laxity of his or her behavior. The owner should then perform a sufficient number of loving acts to redistribute the overflowing pieces to the other player. The level of the game pieces in the containers can be examined at any time to remind the players about whether they have been generous or remiss. The inventor suggests that the game promotes harmony, tranquillity, and peace of mind, is morally uplifting, and is a continuing source of love and pride.

Health and Hygiene

No form of human hygiene is too obscure to escape the notice of inventors trying to bring a little more cleanliness into our lives. Hardly a week goes by without the appearance of three or four new patents for high-tech toilets, but even the lowly bar of soap is constantly being improved. Inventors have probably put more brain power into finding relief from headaches than into curing all the other maladies plaguing mankind. Nontraditional medicine is also well represented as we search for alternative ways to understand our bodies and feel better about ourselves. Surprisingly, patent activity in the field of spittoon design is still going strong.

Abdominal Muscle Firmness Alarm

Patent Number: 4,801,921
Date of Patent: Jan. 31, 1989
Inventor: Robert W. Zigenfus, Evansville, IN

This abdominal muscle firmness alarm consists of an adjustable timer circuit connected to a pair of disk electrodes. The disks are affixed to the user's abdominal region, using tape or an adhesive. The disks deliver either an electrical charge or a vibration to the abdomen when they are energized at regular intervals by the timer. When the user senses the signal delivered by the disks, he or she is reminded to contract the muscles of the abdomen. This increases the muscle tone of the abdomen and serves to flatten it.

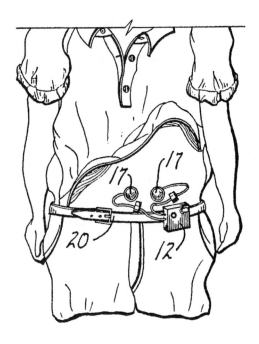

Mouth Appliance for Assisting in Weight Control

Patent Number: 4,883,072
Date of Patent: Nov. 28, 1989
Inventor: Edward W. Bessler, Fort Mitchell, KY

This disposable appliance is designed to inhibit a user's intake of food. Two-pressure sensitive adhesive pads are applied at the corners of the user's mouth. Several lightweight bands connect the pads and arch over the lips of the user without touching them. The appliance permits normal breathing and speech, while at the same time inhibiting the intake of food through the lips. The inventor states that the appliance is less conspicuous than previous food intake restriction devices.

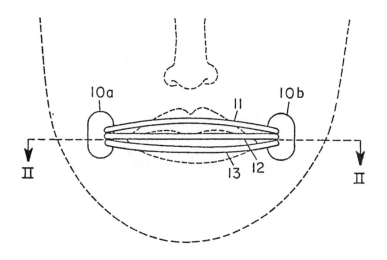

Device for Testing One's Breath

Patent Number: 4,922,921
Date of Patent: May 8, 1990
Inventor: **Laurence B. Donoghue**, Palos Verdes, CA

Individuals suffering from malodorous breath are normally unaware of the situation and may offend those around them without realizing it. This breath tester allows a user to check his or her breath in a quick and simple manner. The tester consists of a rubber or plastic mask that may be held over the nose and mouth. The user takes a deep breath and exhales through the mouth into the mask. Any fresh air initially inside the mask is expelled through the small openings near the top. This leaves the mask full of exhaled gases, which are heavier than air and remain in the bulbous lower portion as long as the mask is held to the face. The user then inhales through the nose, allowing a sample of exhaled breath gases to be evaluated for any noxious odor.

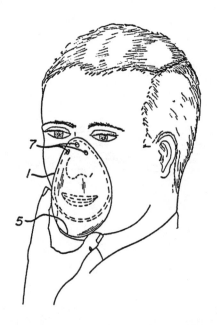

Tooth Storage Means

Patent Number: 4,775,318
Date of Patent: Oct. 4, 1988
Inventor: Daniel V. Breslin, Philadelphia, PA

This child's lost tooth storage receptacle is designed to have the appearance and coloration of the mouth, lips, and gums of a human being. The upper and lower jaws each have a full set of ten imitation teeth, corresponding to those normally found in a child's mouth. Each imitation tooth is hollow and is molded from a transparent plastic. When a tooth is lost by the child, it is deposited into the corresponding hollow imitation tooth. The date of deposit is recorded in the appropriate position on the cover plate. Eventually, the child's full set of primary teeth will be stored within the receptacle and preserved in the proper order. Since the teeth are all clearly visible, they need not be removed for viewing, reducing the probability of loss, damage, or interchange. The receptacle can also be modified for the storage of adult or animal teeth as well.

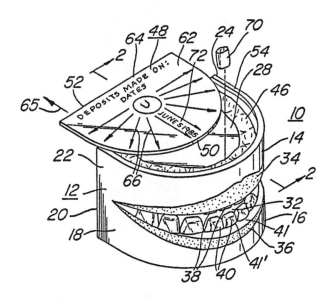

Dental Patient Face and Neck Shield

Patent Number: 4,969,473
Date of Patent: Nov. 13, 1990
Inventor: Susan F. Bothwell, Atlanta, GA

This disposable face and neck shield is intended to be worn by a dental patient during a tooth-cleaning procedure. Its function is to protect the patient's eyes, hair, and upper respiratory system from the mist of fine droplets leaving the mouth when a jet of abrasive solution is used to clean and polish the teeth. It is formed of a lightweight, flexible material that completely covers the patient's face except for the mouth. The mouth opening is surrounded by a pressure-sensitive adhesive coating that seals it to the skin around the lips. This also aids in holding the patient's mouth wide open during the procedure. An air-permeable panel is provided over the nose to facilitate breathing and flexible transparent plastic windows protect the eyes.

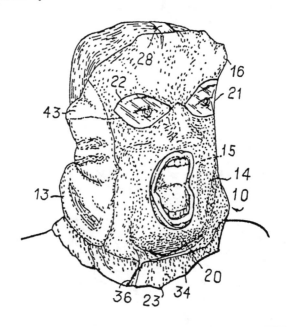

Toothbrush Having Signal-Producing Means

Patent Number: 4,788,734
Date of Patent: Dec. 6, 1988
Inventor: Gerfried Bauer, Richterswil, Switzerland

It is suggested by the inventor that this toothbrush will make it easier for children and adults to brush their teeth for at least a minimum time period. When the toothbrush is gripped, the user's thumb activates a touch switch in the handle. The switch is part of a miniature electronic module that emits a melody or spoken words. The emission of the sound continues for two minutes, the time span regarded by dental science as being the optimum for brushing after each main meal. Battery life is expected to be about three months, at which time the toothbrush should be due for replacement anyway. Since a different melody may be programmed into each toothbrush during manufacture, the acquisition of a new toothbrush affords the user the opportunity to choose a new musical selection.

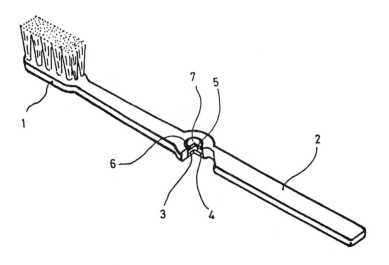

Dental Analgesia Method and Apparatus

Patent Number: 4,782,837
Date of Patent: Nov. 8, 1988
Inventor: Dennis E. Hogan, Hopkins, MN

This invention provides drug-free pain relief during dental procedures by using electronic nerve stimulation. One electrode (*18*) is applied above a nerve location on the patient's hand. A second electrode (*18'*) is placed above a nerve location on the patient's jaw, outside the spot where the procedure is to be performed. When electronic stimulation is applied to these electrodes, the normal response of the nerves is disrupted and the sensation of pain is suppressed. The amount of stimulation received by the nerves is controlled by the patient by means of a slide adjustment mounted on a hand-held control unit (*34*).

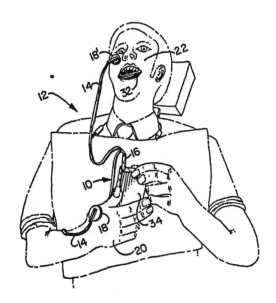

Muscle-Powered Shower

Patent Number: 4,829,609
Date of Patent: May 16, 1989
Inventor: Ernst Debrunner, Zurich, Switzerland

This muscle-powered shower avoids some of the shortcomings of earlier designs and is able to supply the user with a relatively strong stream of water. Two foot-operated bellows (*5, 6*) are used to draw water from an external source (*7*). The user operates these by rocking from side to side or alternating steps on the bellows. Water flows under comparatively low pressure through a flexible plastic tube to the shower head (*2*). It enters the shower head through a tapered channel (*30*) that increases the velocity of the stream. Very little energy is lost by the water stream as it follows the contour of a bell-shaped interior chamber (*31*). The water then exits through holes (*34*) in the lower plate as a comparatively high-pressure shower stream.

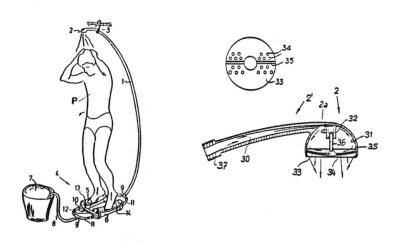

Body Scrubber

Patent Number: 4,817,227
Date of Patent: Apr. 4, 1989
Inventor: John H. Scott, St. Louis, MO

Some people find it difficult to reach certain portions of the body while bathing. This user-powered body scrubber for a tub or shower allows the back and lower extremities to be easily cleaned. An elongated upright guide bar (*14*) is attached to the wall by means of suction cups (*27*) and serves as a track for a pair of movable scrub brushes. The brush carriage (*16*) slides up and down the track as a user alternately pulls and releases a handle (*18*) at the end of a cord or cable (*19*). The brush carriage also contains gears or friction rollers that cause the soft circular brushes to rotate as the carriage moves. A fluid reservoir (*36*) holds a quantity of body lotion, detergent, shampoo, or bath salts that may be introduced into the water stream and sprayed onto the brushes through a set of nozzles (*30*).

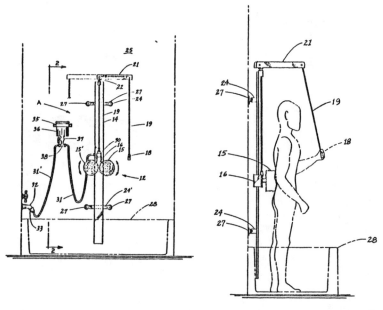

Soap Bar Having a Fifth Finger–Gripping Member

Patent Number: 5,071,583
Date of Patent: Dec. 10, 1991
Inventor: Steve Martell, Chicago, IL

In the past, soap has been formed into a variety of different shapes, with hooks or ropes included to hang it up when not in use. None of these, however, have included any specific way of preventing an ordinary bar-shaped cake of soap from slipping out of a user's hand. This soap bar includes a molded-in gripping device that permits secure retention of the soap while not interfering with scrubbing action. While the thumb and three fingers surround the bar, the user's fifth (pinkie) finger passes through a hole in the gripping device, restricting lateral slippage.

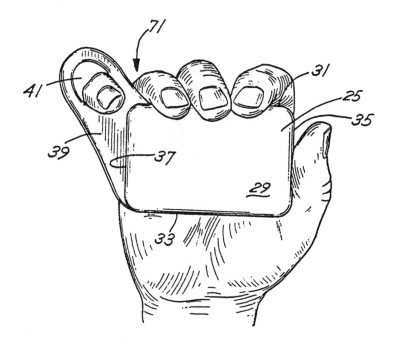

Body Dryer

Patent Number: 5,007,182
Date of Patent: Apr. 16, 1991
Inventors: Sam Fishman, Marlboro, NJ
        Michael J. Marchese, Bayonet Point, FL

The user of this body dryer stands on a platform made of a rigid insulating material such as molded plastic. The rear of the platform is parallel to the ground and then curves upward toward the front. This has the effect of bending the user's toes upward and spreading them laterally, allowing the spaces between the toes to be properly dried to avoid athlete's foot and other fungal diseases. An electric fan mounted underneath the platform blows a stream of air up through a number of apertures in the platform's surface to dry the user's body. No mention is made of any means to heat the flowing air. The inventors suggest that this body dryer will eliminate the need for towels after bathing and may provide enhanced sanitary conditions in health clubs.

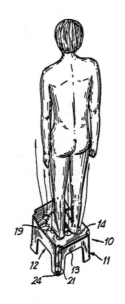

Method for Cleaning Pierced Earlobes

Patent Number: 5,183,461
Date of Patent: Feb. 2, 1993
Inventor: Donna M. Hobbs, Poway, CA

Earring holes are subject to buildup of residue, including dried soap and shampoo, body oils, and skin shed by the tissue surrounding the hole. These residues can collect in the hole, facilitating the growth of bacteria, which can lead to infection. This device is intended to replace previous techniques such as "flossing" to remove earring hole debris. It is constructed from a single strand of approximately 6-inches-long monofilament nylon that is folded in half so that the two ends meet and are parallel. The ends are bonded by heating and melting the strands to form a single stem, about 1.5 inches long. The melting simultaneously creates a rounded tip at the end of the stem which is easily fed through a pierced ear hole. The unbonded portion of the strand forms a loop that is compressed as the stem is pulled out of the opposite side of the ear. The resilience of the loop keeps it slightly expanded so that it gently scrapes the ear hole clean as it passes through.

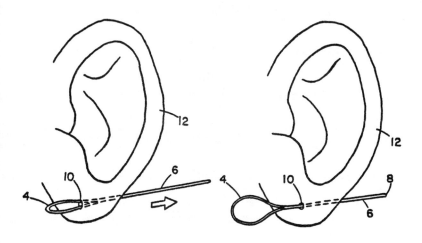

Remotely Actuated Toenail Clipper

Patent Number: 4,847,994
Date of Patent: July 18, 1989
Inventor: Sam Dunn, Jr., Clute, TX

Many older, overweight, or disabled individuals have a great deal of difficulty trimming their toenails with conventional toenail clippers. This improved clipper uses a scissor mechanism to move a plunger rod (*28*) located inside a tubular housing. Squeezing the scissor handles depresses the rod and actuates the lever arm (*13*) of a leaf spring toenail clipper mounted 18 to 24 inches away. An individual seated in a chair can use the toenail clipper easily without excessive bending or stooping. In another version of the invention, the toenail clipper is activated by an electric solenoid and a telescope is included to provide a clear view of the toenail-trimming operation to a user who has poor eyesight.

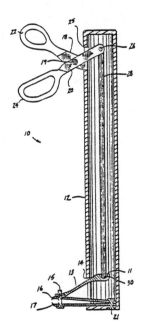

Grooming Aid for Collecting Debris

Patent Number: 4,932,098
Date of Patent: June 12, 1990
Inventor: Richard F. Haines, Los Altos, CA

The ability to maintain the morale of space flight crews is of great importance if missions are to be performed in a professional and efficient manner. This invention allows a crew to perform personal grooming tasks such as hair cutting, shaving, and manicuring under zero gravity conditions and therefore would help to sustain their morale on extended missions. Its basic structure consists of a collapsible ribbed dome covered by a thin, transparent plastic skin. The astronaut to be groomed places his or her head inside the dome and seals a rubber or drawstring-type neck flap. A vacuum hose (*40*) maintains a slight negative pressure inside the dome to collect debris (hair, skin flakes, aerosol droplets, etc.) and prevent its escape into the cabin area. Filtered breathing air is provided through a hose (*44*) and mouthpiece (*46*). The astronaut acting as the barber reaches in through self-sealing arm slits (*55*) near the top of the dome. Self-grooming activities such as shaving can be performed by inserting the arms through slits near the bottom (*53*). Both the astronaut and the barber can be secured to the cabin by placing their feet through Velcro-mounted stirrups (*67*).

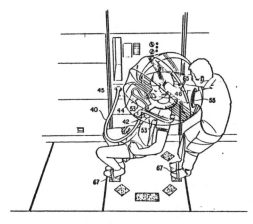

Article of Clothing for Use as a Condom

Patent Number: 4,942,885
Date of Patent: July 24, 1990
Inventors: Anton Davis, Jamaica, NY
Kelvin A. Simmons, Massapequa, NY
Richard Blair, Brooklyn, NY

This invention prevents the slippage of a condom during use. It also provides a greater degree of lower body isolation than does a conventional condom. The body portion is composed of a brief that is held in place by a waist-encircling band. The rim of the condom is securely sandwiched between two mounting plates (*60, 62*). Mating snap fasteners (*56, 70*) on the mounting plate assembly and the brief allow quick installation of the condom when needed.

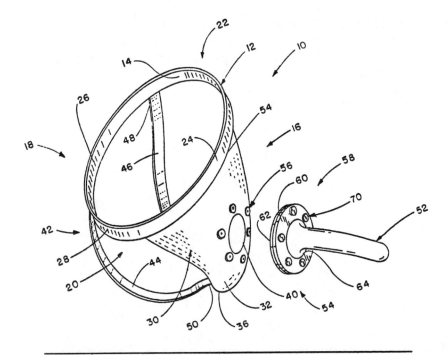

Phosphorescent Prophylactic

Patent Number: 4,920,983
Date of Patent: May 1, 1990
Inventors: Francisco G. Jimenez, San Juan, Puerto Rico
George Spector, New York, NY

This phosphorescent condom consists of a pair of latex rubber sheaths (*12*), one inside the other. A bead of phosphorescent liquid (*24*) is secured between the sheaths at the center of the rounded tip. When the assembly is pulled on over a regular condom (*26*), the resulting pressure breaks the bead and causes the phosphorescent liquid to spread out and completely fill the space between the sheaths. The phosphorescent liquid is activated by exposure to light. After the incident radiation is removed, it gives off a persistent glow across the entire length of the sheaths. No mention is made as to how long the glow lasts.

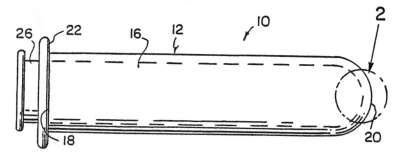

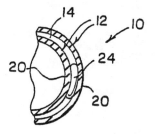

Conductive Condom

Patent Number: 4,971,071
Date of Patent: Nov. 20, 1990
Inventor: Gary D. Johnson, New York, NY

Ordinary condoms are made from electrically nonconductive materials. They block the exchange of any electrical activity that may be present in the nerve endings of the partners and therefore lessen the degree of stimulation perceived. This electrically conductive condom solves that problem. It is made from a thin elastomeric material such as latex rubber. Since rubber is an insulator, the condom also contains numerous microscopic conductive particles made of carbon or silver embedded within it. About one million tiny electrical contacts per square inch are possible. A conductive gel lubricant or spermicide can be used to improve the connection. An optional restraining strap may also be provided. Both male and female versions of the invention have been disclosed.

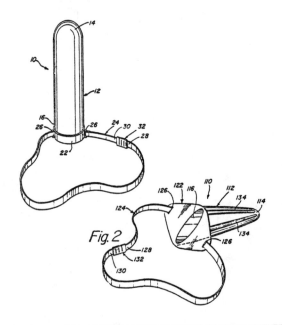

Fig. 2

Vacuum-Assisted Condom Applicator

Patent Number: 4,984,582
Date of Patent: Jan. 15, 1991
Inventors: Gregory Romaniszyn
          Eva Romaniszyn
          both of Fort McMurray, Alberta, Canada

This applicator is intended to solve the problems associated with hurriedly unrolling and possibly damaging a conventional condom. It is constructed from a double-walled concentric tube that is rounded at one end and open on the other. The outer tube (*2b*) is continuous and airtight. The inner tube (*2a*) is perforated by numerous small holes. At the factory, the outer rim (*11*) of a conventional condom (*9*) is placed in a circular groove (*28*) at the open end of the structure. A vacuum is applied to a one-way membrane valve (*7*) located at the tip, causing the condom to adhere to the inner tube wall in a slightly stretched condition. The valve is factory-sealed and the unit is then shipped to the consumer. When the user wishes to apply the condom, a second membrane valve (*6*) near the open end of the structure is depressed. This allows air into the space between the tubes and permits the condom to contract into place by means of its own natural elasticity.

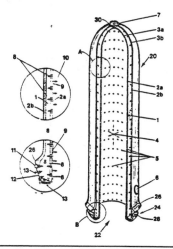

Ball-Point Pen With Condom

Patent Number: 5,007,756
Date of Patent: Apr. 16, 1991
Inventor: Remo C. Wey, La Neuveville, Switzerland

This ball-point pen package permits a person to discretely carry several condoms that are well protected from accidental damage. The condoms (*K8, K9*) are packed into disposable shells (*3, 4*), which may be sold as prepackaged refills. The shells screw together and are closed by means of a cap (*5*) and a conical shell. The conical shell contains a writing fluid reservoir (*6*) and a ball-point pen tip (*1*).

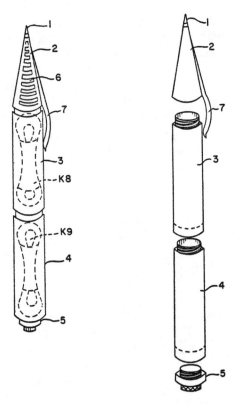

Magnetic Stimulator With Skullcap-Shaped Coil

Patent Number: 5,116,304
Date of Patent: May 26, 1992
Inventor: John A. Cadwell, Kennewick, WA

This invention is intended to provide deep stimulation of the neurons within the cranium in a relatively painless manner. It comprises a coil of heavy-duty wire wound so that its shape resembles a skullcap. The coil is connected to a power supply that produces an alternating high-current electrical wave for a short period of time. A layer of soft material is placed between the coil and the scalp to act as a cushion and to help prevent burns if the coil becomes hot during use. The frequency of the magnetic field produced by the coil is chosen to correspond to the time constant of the neurons to be stimulated. Usable frequencies lie in the range of 3,333 to 20,000 cycles per second.

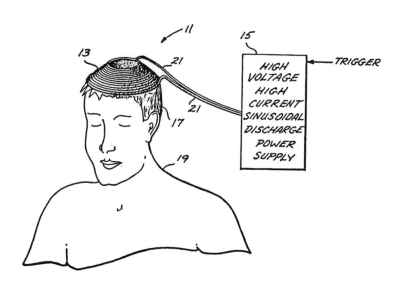

Magnetherapy Insole for Shoes

Patent Number: 5,233,768
Date of Patent: Aug. 10, 1993
Inventor: Clinton C. Humphreys, Plant City, FL

This insole is designed with a magnetic center layer that is sandwiched between cushioning and massaging layers. The magnetic layer has strips of extended permanent magnets laid lengthwise along the insole. Alternate strips are of opposite north and south magnetic polarities. They are separated by a dielectric layer to prevent the strips from becoming neutralized by touching each other over a period of time. The inventor claims that the insole aids in improving the circulation of the blood in the user's feet and legs. It is also suggested that blood is made less acidic, allowing the invention to relieve migraine headaches, even though the magnetherapy insole is worn in the shoe.

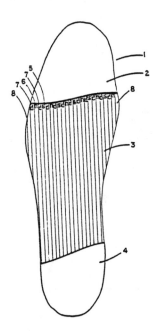

Ultraviolet Germicidal Mask System

Patent Number: 5,165,395
Date of Patent: Nov. 24, 1992
Inventor: Mark R. Ricci, Hingham, MA

This germicidal mask provides a user with biologically uncontaminated air. It could be used by people who are easily susceptible to colds, the flu, or other contagious diseases. Before reaching the inner part of the mask, the air to be breathed passes through a sterilization chamber (3). Inside the chamber, radiation from an ultraviolet light source kills germs, viruses, and other airborne pathogens. A battery pack (6) provides the power necessary to operate the light source. A baffle arrangement prevents any potentially harmful ultraviolet light from directly contacting the skin of the wearer. As an added feature, exhaled air also passes through the sterilization chamber, preventing the user from spreading germs to others.

Headache Treatment Apparatus

Patent Number: 4,781,193
Date of Patent: Nov. 1, 1988
Inventor: Kenneth L. Pagden, Leeton, Australia

The object of this invention is to provide a simple but effective means for the relief or cure of the common headache, migraine, or other head pain. The patient wears an electrically heated cap that raises the temperature of the top of the head. In addition, a headband containing cooling coils circles the patient's brow and temples. The coils are supplied with chilled circulating fluid from a refrigeration unit. According to the inventor, experiments have shown that the combination of heating the crown of the head while cooling a zone just above the ears is effective in relieving a headache.

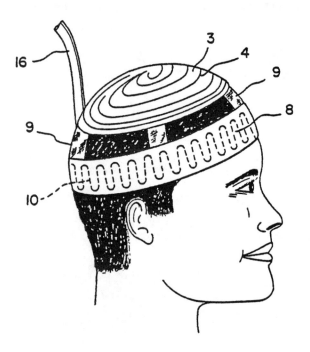

Massaging Head and Face Mask

Patent Number: 4,787,372
Date of Patent: Nov. 29, 1988
Inventor: Keith Y. Ramseyer, Magalia, CA

In this massager, a pair of form-fitting, semirigid head and face pieces are held together by elastic bands. A small, geared-down electric motor (*44*) is mounted on the forehead. The motor drives a crank (*54*) that is connected by a tether to the face piece. When the motor is switched on, the motion of the crank arm repetitively lifts the face piece, massaging the face of the user. The face piece returns to a lowered position in between strokes by a combination of forces due to gravity and facial skin elasticity. By reaction, the headpiece is pulled forward and back during each stroke, massaging the scalp at the same time. The cycle can repeat at a frequency of from forty to seventy strokes per minute. A timer (*48*) ensures that the unit will turn off should the user fall asleep. The inventor suggests that to maintain skin tone, the massager might be used ten to twelve minutes per day for the first two months and every other day thereafter.

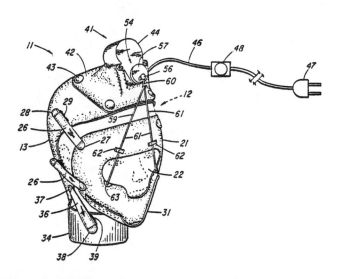

Scalp Massager

Patent Number: 4,807,604
Date of Patent: Feb. 28, 1989
Inventor: Heriberto Canela, Hialeah, FL

Several devices have been designed in the past for applying a pulsating jet of water to different parts of the body. None of these, however, provide the means for applying the massaging jet to a particular area of a person's scalp without requiring constant use of his or her hands. In this invention, water is supplied through a two-way valve assembly (*120*) that permits a user to select either a conventional shower head or the massager. The user wears a helmet containing a number of slots along its length. The massaging jet unit (*40*) is secured at a desired location in a particular slot by tightening a lock nut (*42*). Pulsating water coming out of the massage head (*46*) hits the user's scalp and runs down over his or her body. This way, a user gets not only a scalp massage by also a body rinse.

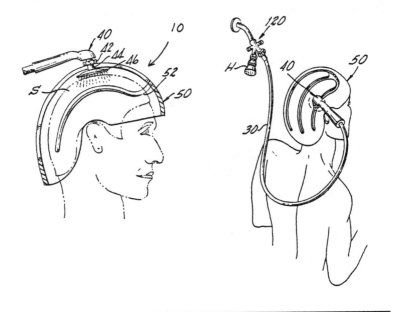

Massaging and Combing Helmet for Insomnia

Patent Number: 5,081,986
Date of Patent: Jan. 21, 1992
Inventor: Nam I. Cho, Ashland, MA

Because head massage induces warm, contented feelings and relieves tension, it is an effective relaxation promoter and a useful treatment for insomnia. Unfortunately, most insomniacs do not have access to a nightly head massage by another person. This massaging helmet solves the problem by providing a massage similar to that given by a human hand. It contains an endless belt made of a soft plastic that has various textures. The belt has both an embossed area (*46*) to mimic the undulations of the human palm and another area with rows of upright fingers (*48*) that comb the hair. A small electric motor (*24*) and rechargeable batteries (*20*) drive the belt from front to back for about thirty minutes on each charge. Brushing the hair backward before using the helmet is recommended to avoid tangling it in the belt.

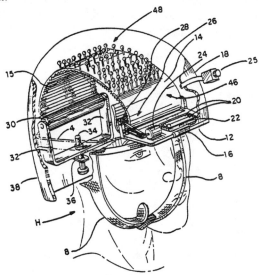

Drug-Free Method for Treatment of the Scalp for Therapeutic Purposes

Patent Number: 5,228,431
Date of Patent: July 20, 1993
Inventor: Ralph R. Giarretto, Oakland, CA

Scalp massages are frequently recommended for balding people under the premise that the massage helps to increase the flow of nutrients and oxygen to hair follicles. This treatment provides an improved method for stimulating the scalp by alternately applying ambient and negative pressures. During the treatment, the user wears a lightweight, rigid helmet. To provide an airtight fit with the user's head, an expandable seal (*15*) inside the helmet is inflated with a hand bulb (*14*). A vacuum source unit (*20*) alternately blows lukewarm air onto the scalp and then reduces the pressure in the helmet, causing blood to rush to the hair follicles. After treatment, the scalp becomes red and warm, oily, and often sweaty. The treatment is administered for several minutes daily and is claimed to result in the prevention of hair loss and the restoration of hair growth. It is also said to be useful for relieving stress-induced migraine and sinus headaches.

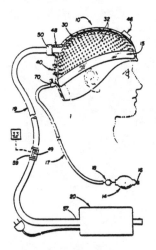

Toilet Bowl Illuminator

Patent Number: 5,136,476
Date of Patent: Aug. 4, 1992
Inventor: Donald E. Horn, Uniontown, OH

Adults needing to use a toilet during the middle of the night often find that switching on the room lights is irritating and disorienting. Extinguishing the room lights afterward creates further disorientation while the eyes become reacclimated to the darkness. Additionally, adults with small children who are incapable of reaching the wall switch themselves often find that they are awakened for assistance in the middle of the night. This battery-powered toilet bowl illuminator hangs from the toilet bowl rim and provides suitable illumination for locating and utilizing the facilities. It may be used with the seat up or down. The light source is a low-power green light-emitting diode (*32*) suspended above the bowl. A photoelectric sensor automatically turns the light source off during the day when sufficient room illumination is available.

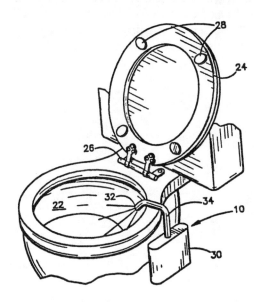

Toilet Seat Cover Position Alarm

Patent Number: 4,849,742
Date of Patent: July 18, 1989
Inventor: Blake Warrington, Dublin, CA

An open toilet bowl does not present a particularly attractive appearance and creates other problems as well. Small children and household pets may be attracted to an open toilet and may attempt to enter the bowl. Also, if both the cover and the toilet seat are left in the raised position, the condition can be highly discomforting to a subsequent user who fails to notice the situation. This invention provides a signal that is activated for a period of time if the toilet seat cover is not promptly lowered following flushing of the toilet. The alarm unit housing (*19*) hangs from the toilet tank by means of a hook-shaped bracket. A float switch (*36*) senses the water level in the tank. A magnetic reed switch (*24*) senses the position of the toilet seat cover in conjunction with a small magnet (*46*) mounted on the cover itself. After the toilet is flushed, the water level drops and closes the float switch. If the seat cover is up, the alarm is activated. The alarm is immediately deactivated when the seat cover is lowered. If the seat cover remains up, the alarm is eventually deactivated by the rising water level in the tank.

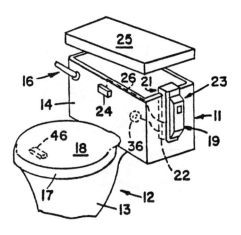

Integrated Passenger Seat and Toilet Apparatus and Method

Patent Number: 4,785,483
Date of Patent: Nov. 22, 1988
Inventor: Paul H. Wise, Tucson, AZ

Vehicles are supplied with toilets to enable passengers to travel long distances without leaving the vehicle, as well as for a variety of other reasons. This invention provides a method for installation of a toilet so that it is integrated into a passenger seat and requires no additional space within the vehicle. The entire structure is mounted on ball bearings so that it swivels like a conventional van seat. In order to use the toilet, a privacy curtain is drawn around the seat and one of the pivoting armrests (*17*) is raised. The hinged seat cushion (*16*) is unlatched (*54*) and then flipped over. When the foot-operated flush lever (not shown) is depressed, an electrically driven pump provides rinse water from a reservoir. Waste is collected in a holding tank beneath the vehicle. A deformable gasket (*16c*) seals the toilet tightly when not in use.

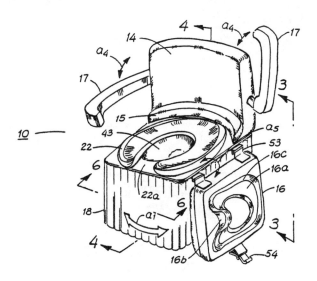

Musical Potty Chair

Patent Number: 4,777,680
Date of Patent: Oct. 18, 1988
Inventor: Lirida Paz, Elizabeth, NJ

This musical potty chair is intended to be used during the toilet training of a small child. Before use, a lightweight plastic liner is fitted into the receptacle (*30*) and the music box key (*72*) is wound. When the child eliminates, the waste is deposited onto the surface of a spring-loaded platform (*52*). This small additional weight compresses the spring (*60*) slightly, allowing the end of the platform support rod (*64*) to depress a lever (*74*). Movement of the lever releases the escapement of the music box (*70*), rewarding the child's efforts with a tune.

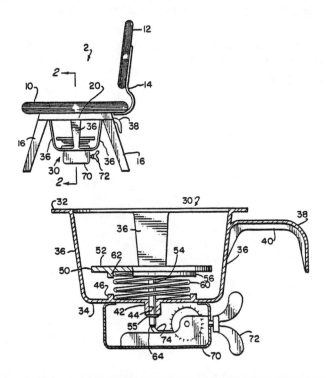

Urinal for Use by Female Individuals

Patent Number: 4,985,940
Date of Patent: Jan. 22, 1991
Inventor: Kathie K. Jones, Pensacola, FL

Since healthy individuals of both sexes urinate in about the same length of time, the long lines at women's rest rooms are primarily due to the absence of convenient plumbing fixtures of the type commonly referred to as "urinals." This invention consists of a plumbing fixture for installation in women's rooms that enables female individuals to urinate from a standing position. Its function and manner of use are intended to be readily understood even by someone who sees it for the first time. The ceramic fixture contains a water-filled basin (*18*) that is just above floor level. An elongated flexible hose (*26*) has a urine-collecting funnel (*22*) mounted at the top and a lower end that feeds into the basin. The funnel is removed from its hanger (*30A, 30B*) by means of a handle (*24*) and is used to engage a sanitary cuff from a wall-mounted dispenser (*70*). The sanitary cuff lines the rim of the funnel so that the funnel does not come into contact with the body of the user. The cuff is automatically ejected when the funnel is replaced in the hanger. A wall-mounted flush button rinses the bowl and disposes of waste.

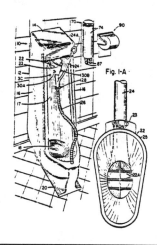

Fig. I-A

Portable Pocket Spittoon

Patent Number: 4,885,809
Date of Patent: Dec. 12, 1989
Inventor: Charles H. Muchmore, St. Bernard, OH

This portable pocket spittoon is intended to permit the tobacco chewer to discretely dispose of excessive fluid without publicly displaying an unsightly stream of tobacco juice. The spittoon is constructed in the form of a plastic flesh-colored bottle. It is intended to be used with one hand. A unique molded-in grip allows the user to retain it securely. A thumb-operated hinged lid allows easy access, but helps to prevent spilling the contents. The open lid and index finger of the user's hand additionally serve to screen the user's mouth from the view of bystanders when fluid is discharged into the container.

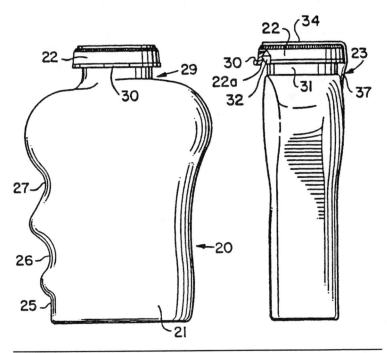

Flushable Vehicle Spittoon

Patent Number: 4,989,275
Date of Patent: Feb. 5, 1991
Inventor: Dan L. Fain, Chancellor, AL

This flushable spittoon is intended for use by a driver or passenger of a motor vehicle. The user deposits fluid into a cylindrical receptacle (26) that is mounted by means of a Velcro pad at a convenient location within the vehicle. Under the action of gravity, the fluid flows downward into the funnel-shaped bottom of the receptacle, enters a flexible drainage tube (36), and is finally discharged onto the ground under the vehicle. The user may also activate a small toggle switch mounted on the side of the receptacle. This causes windshield washer fluid to be drawn from the vehicle's reservoir (16) by a supplementary electric motor-powered pump (52). The fluid is then conveyed by a tube to a spray jet nozzle that rinses the interior of the spittoon receptacle.

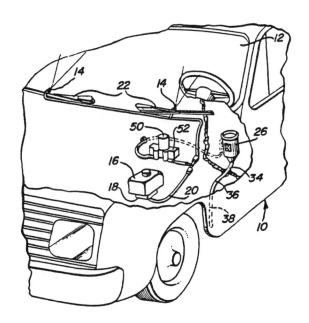

Devices for Aiding Resynchronization of Body Clocks

Patent Number: 4,893,291
Date of Patent: Jan. 9, 1990
Inventors: Peter A. Bick, Fife, Scotland
            Christine J. Kinnell, Edinburgh, Scotland

This invention provides a computer-controlled display that indicates a procedure to resynchronize a user's biological clock after a plane flight across many time zones. It operates on the principle that a controlled exposure of a traveler's body to daylight can advance or retard the body's internal clock. The user enters the departure time, flight duration, local arrival time, and flight direction (east or west). Upon arrival at the destination, the display provides instructions as to whether to seek out or avoid sunlight for a specific number of hours.

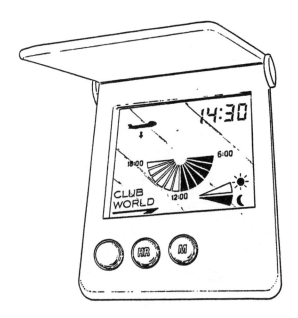

Mood-Altering Device

Patent Number: 4,777,937
Date of Patent: Oct. 18, 1988
Inventors: Charles Rush, Peter J. Guinan, both of Woodstock, NY

This invention utilizes a combination of visual and auditory stimuli for treating stress and enhancing relaxation. The user wears a facial mask that excludes external light and presents the eyes with a continuous glowing light of uniform color and texture. The uniform light is produced by an electroluminescent panel inside the mask. This induces an effect known as ganzfeld. Ganzfeld is a condition whereby a subject loses all sense of visual perception. It is a complete loss of "seeing" in that the subject does not even know whether his or her eyes are open or closed. Simultaneously with the light, a gentle rhythmic pink noise is introduced to the ears through a pair of earphones (18). Pink noise simulates the sound of flowing water. The ganzfeld effect combined with the monotonous rhythmic sound produces a powerful mood alteration in the individual. Stress is quickly reduced and the user becomes relaxed and serene. The user may select different colors for the visual display from a control panel, but blue or green are claimed to be the most useful in enhancing relaxation and relieving stress.

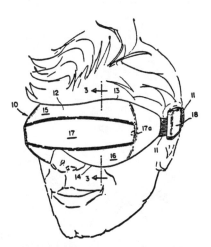

Character Assessment Method

Patent Number: 5,190,458
Date of Patent: Mar. 2, 1993
Inventor: Valma R. Driesener, Maylands, Australia

This invention provides a psychological test for character assessment, derived from a drawing done by an individual. The person is instructed to draw a picture that incorporates a number of interconnected graphic symbols. The symbols (and their meanings) are selected from the following group: a hand (self-image), an eye (authority), a tree (life history), a fish (sexuality), a star (none listed), a spiral (progression or development), a half circle (shelter or openness) and a zigzag (stress). Further symbols that also may be used are a flower (creativity), a sun (warm family support), a bird (freedom), and a wave (change). Interpretation of the drawing is made based on the relation between the symbols. For example, a large eye drawn on a fish might mean one's sexual partner is in a position of authority. Since the meanings of the symbols are not readily known by a subject, falsification of the results is difficult. One significant factor in a drawing is the direction in which a symbol is drawn, so that it is desirable to keep a record of where the symbol was started and finished.

Hand Reflexology Glove

Patent Number: 5,199,876
Date of Patent: Apr. 6, 1993
Inventor: Martin S. Waldman, Vandenberg, CA

Reflexology is the practice of using direct pressure to stimulate the reflex receptor points in the hand. These points directly correspond to the body's internal organs and functions. The body is divided into ten zones of jurisdiction, each zone containing a specific set of organs with their reflex counterparts in the hand. This invention consists of a pliant plastic glove that is imprinted with a reflexology zone map. It enables a novice practitioner to benefit from a deep, focused, and penetrating self-application of hand reflexology massage. The glove also shields the user from the distracting effects of skin friction as well as the skin's sense of touch. It promotes heating and perspiration of the hand and may be used to contain special massage lotions. The inventor suggests that use of the glove quickly trains an individual to be aware of messages sent to the reflexes in the hands, which may signal the possibility of an impending illness.

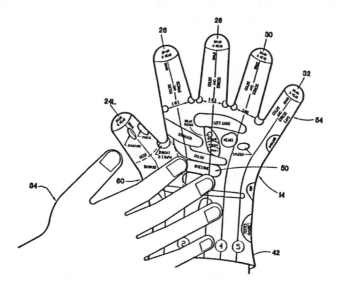

Bi-Digital O-Ring Test for Imaging and Diagnosis of Internal Organs of a Patient

Patent Number: 5,188,107
Date of Patent: Feb. 23, 1993
Inventor: **Yoshiaki Omura**, New York, NY

This method is claimed to accurately image an internal organ of the body without the use of X rays or other potentially hazardous procedures. The patient forms an O-ring shape with one hand by placing the tips of the thumb and another finger together. A sample of tissue of the internal organ to be imaged is placed on the patient's other hand. This sample may be a microscope slide preparation and does not to have come from the patient's own body. It may in fact be any mammal organ, such as that of a cow, a monkey, or a pig. The patient's own internal organ is then externally probed with a rod-shaped instrument. Simultaneously, a tester attempts to pull apart the O-ring shape. According to the inventor, the electromagnetic field of the tissue sample interacts with the electromagnetic field of the internal organ of the patient being probed. This results in an interaction that allows the O-ring shape to be pulled apart when the boundaries of the organ are reached by the probe. The force required to pull the O-ring shape apart is related to the health of the organ at the location of the probe.

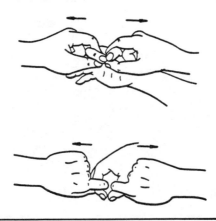

Novelties

We all love being surprised by something new and unusual. In this section, we find some inventions designed to be just a little bit different.

Life Expectancy Timepiece

Patent Number: 5,031,161
Date of Patent: July 9, 1991
Inventor: David Kendrick, Berkshire, NY

This timepiece is used to monitor and display the approximate time remaining in a user's life. A microprocessor monitors the passage of time. A resettable memory is connected to the processor for storing data representative of years, days, hours, minutes, and seconds. Buttons or switches are provided to enter and change the stored data so that the approximate time remaining in the user's life can be reset by the user. The user determines his or her life expectancy by referring to a combination of actuarial and health factor tables. In the example shown, the display indicates that the user has 37 years, 110 days, 21 hours, 4 minutes, and 42 seconds left to live.

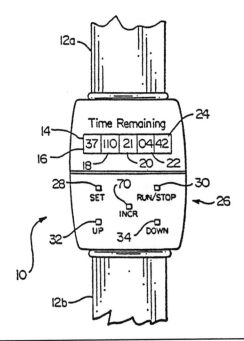

Package

Patent Number: 4,776,511
Date of Patent: Oct. 11, 1988
Inventor: Martine Tischer, Mundelein, IL

Many materials have previously been proposed for wrapping an object to form a gift package. This unique package wrapper is composed of a sheet of uncut negotiable bills of paper currency. The inventor indicates that there are a number of advantages associated with using uncut sheets of money for wrapping paper. The green color of U.S. paper currency is very attractive and suggests power and wealth. The wrapping can also be cut into individual bills and used as negotiable currency, which saves money on ordinary wrapping paper and avoids waste.

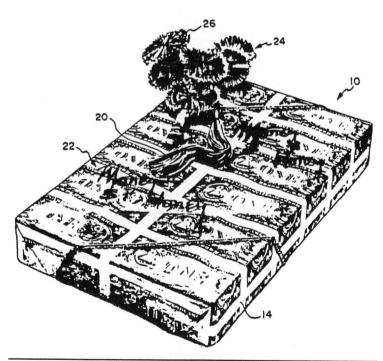

Animal Head

Patent Number: 4,778,172
Date of Patent: Oct. 18, 1988
Inventor: William C. Bryan, Chatsworth, CA

This molded rubber animal head contains a set of open jaws that fit around a person's wrist. An elongated tongue wraps around the wrist and is secured by Velcro. By making the tongue flesh colored and providing red coloration (62) on a visible portion of the tongue, the effect of blood dripping from a bite can be simulated. Since the tongue secures the animal head to the wrist, it can be easily carried about without being held in the hands and interfering with the activities of the wearer.

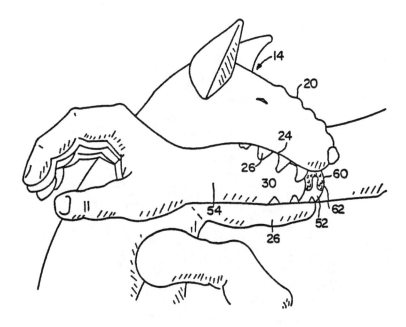

Wigglin' Fish Amusement and Novelty Device

Patent Number: 4,775,351
Date of Patent: Oct. 4, 1988
Inventor: Victor Provenzano, Jr., Ventura, CA

This novelty device is intended to closely simulate the appearance and movements of an actual fish, fresh out of the water. The fish body is made out of rubber or a deformable resilient plastic. A battery-powered motor inside the hollow body cavity drives a flexible rotary shaft (28) that is slightly bent and runs the length of the fish. Attached to the shaft are a number of eccentrically mounted discs (32). The motor may be energized by a manual or sound-activated switch (38). When the shaft rotates, the discs impart a lifelike thrashing motion to the body of the fish. The inventor suggests that it is a particularly appropriate gift for an overzealous fisherman who always seems unable to "bag" the big one.

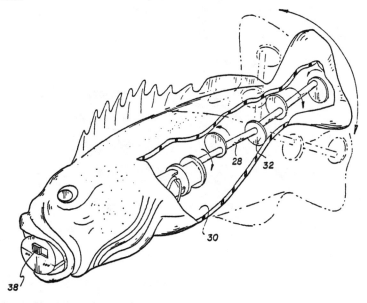

Novelty Soap

Patent Number: 4,861,505
Date of Patent: Aug. 29, 1989
Inventor: Jacqueline Farman, New York, NY

Inside this bar of soap is an electronic circuit module that emits a visible light, tone, musical melody, or message when the bar is picked up. The circuit is activated when a magnetic reed switch in the module is removed from the immediate vicinity of a magnet (*19*) located in the soap dish. Alternatively, a vibration- or temperature-sensing switch may be used. The circuit is enclosed in a water-impermeable housing that is large enough to prevent it from being swallowed by a small child when the soap eventually dissolves.

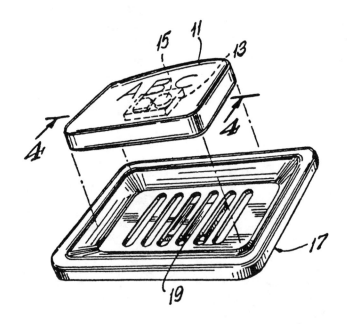

Interest-Paying Bank

Patent Number: 5,135,434
Date of Patent: Aug. 4, 1992
Inventor: Andrew B. Mallon, New York, NY

For many years there has been a need for impressing upon people the benefits of saving money. Currently, the United States has a savings rate lower than that of other industrialized countries. According to the inventor, a toy bank that encourages children to save money while simultaneously teaching the concept of interest would help solve a problem of national importance. When a coin of a given denomination (e.g., a quarter) is deposited into this toy piggy bank, it depresses a sliding component (*93*). The sliding component engages a string (*94*) and pulls a sliding tube (*95*) containing a supply of stacked coins (e.g., nickels) to the right. One coin from the stack falls through a hole (*96*), slides down a ramp (*99*), and emerges as an interest payment through the pig's mouth (*100*).

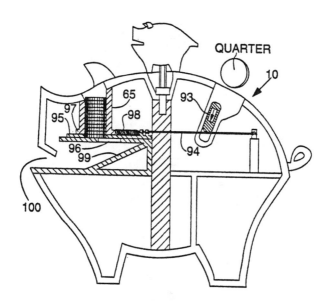

Toilet Seat Clock Apparatus

Patent Number: 5,182,823
Date of Patent: Feb. 2, 1993
Inventor: Ron Alsip, Raynham, MA

Modern families often include two wage earners who each have early morning time pressures, making multiple use of the bathroom and toilet facilities a potential source of friction. This is particularly true when no clock is available in the bathroom for monitoring the time left until departure for work. This waterproof digital clock mounted within a recess in a toilet seat (*24*) provides users with a readily available time reference. The seat cover has a rectangular cutout (*15*), enabling the clock to be viewed even when the seat cover is down. The clock display may be reversed so that the numbers are easily readable by a user either sitting on the seat, or standing facing the toilet. No mention is made regarding the use of an alarm to limit a user's allotted time within the bathroom.

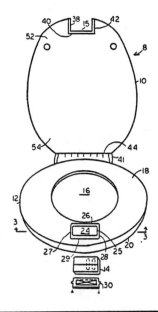

Wrestling Bed

Patent Number: 5,182,824
Date of Patent: Feb. 2, 1993
Inventor: Nickolas A. Cipriano, Philadelphia, PA

This wrestling bed provides a well-defined and restricted area for children's vigorous indoor activities and serves to prevent accidental damage to the rest of the house. The general appearance of the bed lends realism to children's wrestling matches. Its compact size allows for installation in a standard bedroom. The wrestling bed is solidly constructed to resist damage from a level of active play that might harm an ordinary bed frame. It may contain a twin- through king-size resilient mattress, which provides greater protection than a real wrestling ring surface. Peripheral padding and padded corner posts are provided for additional safety. When tired, a child can of course use it as an ordinary sleeping bed.

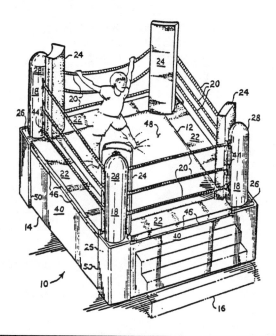

Greeting Card Confetti Delivery System

Patent Number: 4,787,160
Date of Patent: Nov. 29, 1988
Inventor: Lawrence J. Balsamo, Roselle, IL

Several types of greeting cards that spew confetti on the recipient have been invented in the past. Most of them either leak confetti while the card is being removed from the envelope, or cannot be personalized without spilling the contents. Other variations contain the confetti within a pouch, producing a suspicious bulge and typically resulting in a slow, cautious opening of the card. This design overcomes the limitations of previous models. The confetti is contained in a thin paper packet (*20*) housed within a recess in the card. After personalizing the inside of the card, the sender removes the protective layer (*17*) from the rupturing adhesive (*16*) and closes the card. The next time it is opened, the packet is torn and the confetti is sucked out by the resulting air currents. The confetti used in the packet may include a variety of finely divided materials, such as paper, plastic film, metal foils, and even seeds and spices. The type of confetti may differ depending on the inscription on the card. For example, in a card suggesting that the giver is presenting the recipient with a large amount of money, the packet may contain shredded currency as confetti.

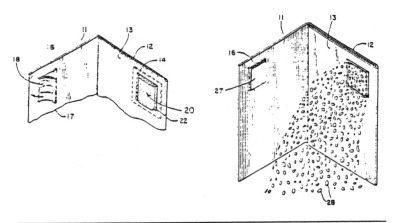

Sandal

Patent Number: D. 308,125 (Design Patent)
Date of Patent: May 29, 1990
Inventor: Peter I. Fraser, Eugene, OR

This unique sandal is sure to attract attention wherever it is worn.

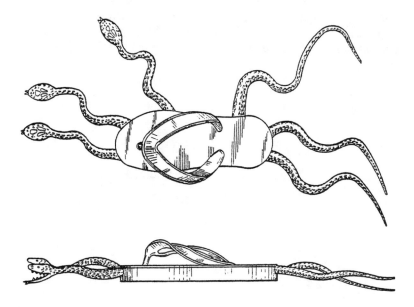

Flip Bill Cap

Patent Number: 4,777,667
Date of Patent: Oct. 18, 1988
Inventors: Barton H. Patterson, Sioux Falls, SD
George Spector, New York, NY

This novelty cap has a flip-up bill with an entertaining advertising message written on the underside. The bill is hinged (*16*) and is normally held in a lowered position by means of a spring (*30*). A long cord (*36*) runs from a pull ring (*40*) through a number of eyelets (*34*) and terminates on an internal arm (*28*) connected to the bill. When the cord is pulled, the bill flips upward. In another version of the invention, a switch activates a battery-operated solenoid to flip the bill.

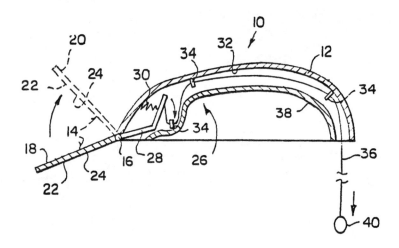

PETS

Our pets are an extension of ourselves, so it's not surprising that inventors would look for ways to improve their lives along with our own. In this section, we see inventions designed to help our pets be cleaner, happier, and more comfortable. Patents covering technology to take some of the work out of pet ownership are represented here as well.

Dog Umbrella

Patent Number: D. 324,117 (Design Patent)
Date of Patent: Feb. 18, 1992
Inventor: Celess Antoine, Forestville, MD

This patent covers the design of a self-supported umbrella to be carried by a dog. The base of the umbrella is fixed to the dog's midsection with a wide band. It is made of a transparent material to allow the dog to readily see out. Air holes are provided near the dog's nose to provide oxygen and prevent fogging.

Electrically Lighted Leash

Patent Number: 4,887,552
Date of Patent: Dec. 19, 1989
Inventor: James T. Hayden, Cincinnati, OH

This improved dog leash is illuminated along its length by a series of closely spaced internal lights. The lights provide a nearly continuous "line" of light between the owner and the pet, which improves their visibility from a distance and contributes to safety. The leash is constructed from a transparent hollow flexible tube with a thin rib running down each side. A number of 9 volt bulbs, spaced from 1 to 6 inches apart and wired together in a parallel line, are located in the tube's hollow core. The hand loop and choker-type collar are illuminated as well. A battery pack (*30*) and switch (*31*) are used to energize the lights.

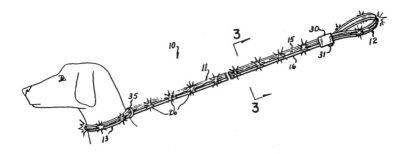

Sleeping Bag for Pets

Patent Number: 4,893,586
Date of Patent: Jan. 16, 1990
Inventor: Betty J. Carson, Canyonville, OR

Individuals engaged in outdoor activities such as hunting, camping, and hiking frequently desire to take a pet along on the outing. Unfortunately, they are prevented from doing so by a lack of adequate shelter for the pet. Rigid shelters are obviously highly inconvenient and impractical when traveling with a pet. Another problem encountered by pet owners is providing adequate warmth to permit the pet to remain out-of-doors during the winter months. This pet sleeping bag is shaped to appeal to the natural tendency of a dog or cat to crawl or "burrow" into a shelter. A permanent arched opening allows the animal to easily find the entrance and exit and avoids the risk of suffocation. Grommet holes are provided to stake down the bag when used out-of-doors. The pet sleeping bag has no rigid components and so may be easily rolled up for transportation.

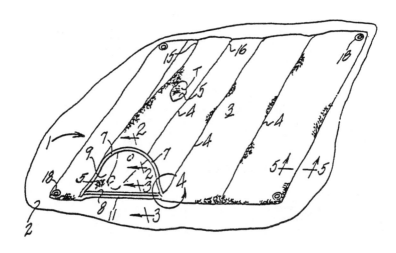

Animal Hat Apparatus and Method

Patent Number: 4,969,317
Date of Patent: Nov. 13, 1990
Inventor: April Ode, Lake Havasu City, AZ

This hat is intended to protect a four-legged animal from the harmful effects of extreme heat and direct sunlight. The brim protects the animal's eyes, face, ears, and rear neck. A Velcro chin strap holds it in place. The hat also contains an enclosed cavity filled with a spongelike absorbent material. The absorbent material may be saturated with a cool liquid that slowly seeps downward onto the animal's head. This provides evaporative cooling and helps to keep the animal at a comfortable and healthy temperature over a relatively long period of time.

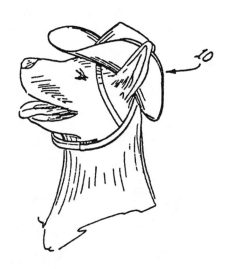

Throw and Fetch Doggie Toy

Patent Number: 4,995,374
Date of Patent: Feb. 26, 1991
Inventor: William L. Black, Margate, FL

As any dog owner knows, some dogs never get enough of the old throw and fetch game. The human owner usually wears out long before the dog is ready to call it a day. Furthermore, many pet owners suffer from arm ailments, lack of muscular strength, and old age so that they cannot exercise their dogs in this manner at all. This ball-throwing apparatus enables a dog to play throw and fetch unassisted for as long a period of time as it desires, regardless of the physical capability of the owner. The ball is launched out of a cylindrical barrel on an airborne trajectory by an electric solenoid. The azimuth and elevation of the barrel are adjustable. The entire upper surface of the invention's enclosure is funnel-shaped, providing a relatively large collection area to make it easy for the dog to feed the ball back inside. A sensor automatically triggers the launching solenoid each time the ball returns to position.

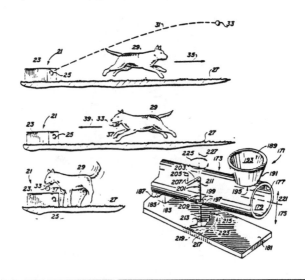

Animal Exercising Device

Patent Number: 4,777,910
Date of Patent: Oct. 18, 1988
Inventor: Francis H. Pecor, Greenville, SC

Due to certain injuries, diseases, and old age, the rear legs of an animal may become incapacitated and not receive adequate exercise during its normal course of activity. Such a condition prevents one of the major muscle groups from being actively stimulated, resulting in diminished overall well-being and physical conditioning of the animal. This invention allows an animal's disabled pair of rear legs to be exercised as it walks normally on its front legs. The device consists of a pair of spoked wheels connected by a U-shaped frame (*12*). The frame is padded (*30*) where it crosses the animal's back. Each leg attachment assembly (*16*) includes a pliable plastic boot for receiving a rear paw of the animal. The animal's rear legs are exercised as the rear wheels turn and the attached crank arms oscillate up and down.

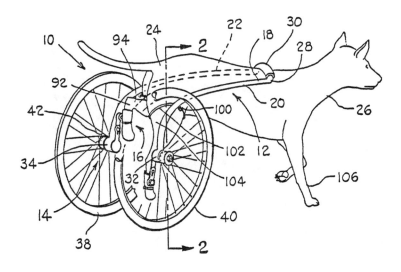

Animal Retentive Fence Attachment

Patent Number: 5,143,354
Date of Patent: Sep. 1, 1992
Inventor: McEdward M. Nolan, Birmingham, AL

Many dogs escape from fenced yards by a combination of jumping and climbing. They are agile enough to leap to the top of the fence and cling to the top rail with their forelimbs, while using their rear feet to push themselves over. Some dogs do not jump at all. They climb from the bottom as if on a ladder and may be injured due to falls while attempting this escape. This invention is intended to keep the clever canine confined safely. It consists of a fence topping that is unstable and will not support the weight of a dog attempting to go from one side to the other. The fence is topped by flexible plastic arms (*41*) that bend back toward a dog trying to climb over. The arms spring back to an extended position after the dog slides off and falls away from the fence. In another form of the invention, the fence top is covered with rollers that prevent the dog from obtaining a secure grip.

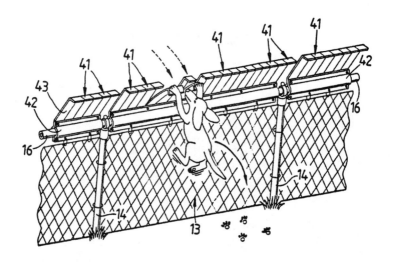

Scratching and Petting Device for Household Pets

Patent Number: 4,872,422
Date of Patent: Oct. 10, 1989
Inventor: Rita A. Della Vecchia, Bend, OR

Prior inventions for stroking or scratching a pet have been stationary objects. Any motion involved had to come from the pet itself and required a fairly high degree of training and intelligence on the part of the pet for proper use. This invention requires only the presence of the pet in the immediate vicinity of the device, with all of the scratching, stroking, and petting being accomplished by the mechanism itself. An electric eye (*60*) determines when a pet is in position under the resilient petting hand (*44*). The electric eye activates a motor and gear train in the petting arm, which cause the hand to slowly pivot up and down about the wrist (*45*). The inventor suggests that the appearance and natural motion of the hand causes the pet to associate this pleasant activity with its human owner. The device may be folded against the wall when not in use.

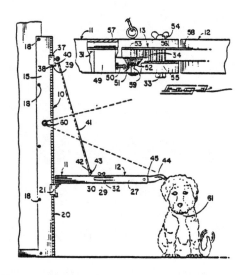

Protective Ear Bags for Dogs

Patent Number: 4,964,264
Date of Patent: Oct. 23, 1990
Inventor: Depy P. Adams, Charlotte, NC

Long-eared dogs such as cocker spaniels and setters commonly soil their ears while eating. Dirty ears are unsightly and quickly develop an unpleasant odor unless frequently washed. This disposable protective covering can be placed on a dog's ears without straps or fasteners and can be removed by the dog itself when it is finished eating. The ear protector consists of a pair of "ZipLoc" transparent plastic sandwich bags connected together by a wire twist tie. The ear bags are not tied to the dog in any way. They are merely suspended from the top of the dog's head, with the weight of the ears themselves holding the device in place. When the dog has finished eating, the ear bag set may be removed, or the dog may be trained to shake it off. The inventor credits a white and brown cocker spaniel named Annie for providing the essential motivation for the conception and reduction to practice of this invention.

Device With Pouches for Receiving Animal Waste

Patent Number: 5,146,874
Date of Patent: Sep. 15, 1992
Inventor: Stella M. Vidal, New York, NY

In urban areas, sidewalks, curbs, alleyways, parks, lawns, and yards are constantly soiled by pet wastes, making life miserable for pedestrians who inadvertently step in the stuff. There are also indoor public places that are subject to the same problem and pet owners often find that rugs and furniture in the home are ruined by their pets. This invention provides a number of disposable pouches to receive and retain the body wastes of dogs, cats, and other mammals. It adapts to the anatomy of the wearer to allow a high degree of comfort and a sleeve surrounds the tail to prevent leakage. Many variations of the device are depicted (in 155 drawings) in the patent disclosure corresponding to male and female animal anatomy.

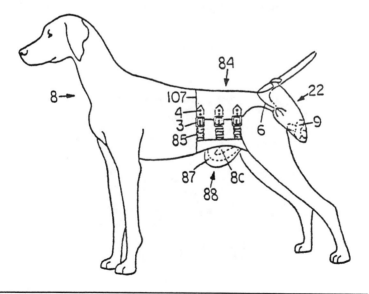

Pet-Grooming Restraint Device

Patent Number: 5,243,931
Date of Patent: Sep. 14, 1993
Inventor: Richard W. McDonough, Cincinnati, OH

When pets are bathed, soap and water are usually splashed everywhere. Compounding this problem, dogs and cats have a tendency to shake off excess water before they can be fully dried, further spraying water on the person bathing the pet and the surroundings. In addition, some pets absolutely refuse to stand still for any kind of grooming activity. This invention provides an enclosed pet-bathing container that simultaneously restrains the pet and contains the mess. The bottom of the container is dimensioned to fit into a standard bathtub and casters also facilitate its use for transporting the pet. A removable lid with air holes (*105*) seals the enclosure. A harness restricts vertical and forward-backward movement, while allowing the pet relatively unrestricted side-to-side freedom. The groomer's arms enter the container through self-sealing slits (*30*) and operate a shower head (*60*) connected to an external water supply. Water drains out the bottom when a stopper (*71*) is removed.

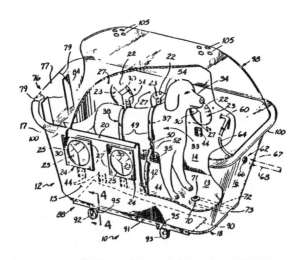

Safety and Security

Inventors can be shrewd tacticians in the endless war on crime. In this section, we see how they supply us with novel countermeasures to protect us against assault and theft from equally inventive criminals. Several accident prevention applications are also illustrated.

Collapsible Riding Companion

Patent Number: 5,035,072
Date of Patent: July 30, 1991
Inventor: Rayma E. Rich, Las Vegas, NV

This invention provides a collapsible riding companion as a criminal deterrent. It includes a simulated human head and torso maintained upright by draping a connected piece of fabric over the car seat back. A series of horizontal metal weights sewn into the fabric provide added stability. The torso may be covered with a T-shirt or a zippered collared shirt. When not in use, the simulated human head retracts into the torso section and the weighted fabric is removed from the seat back to cover the torso. In this form, the riding companion becomes a lightweight, easy-to-carry rectangular case for traveling.

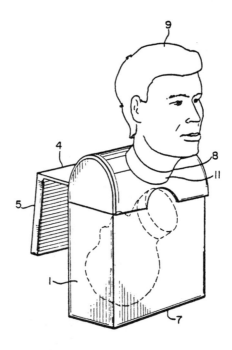

Surveillance System Having a Miniature Television Camera Mounted Behind an Eyeball of a Mannequin

Patent Number: 4,982,281
Date of Patent: Jan. 1, 1991
Inventor: Frederic J. Gutierrez, Denver, CO

In this invention, the lens of a miniature television camera is disguised as the eye of a mannequin, so that customers will not be aware that the mannequin is actually a video surveillance system. In a later patent (Number 5,111,290), the inventor adds a small transmitter and battery power pack to enable the mannequin to be used, without connecting wires, at any location within the store. Multiple mannequins may be employed to view different areas of the store, each functioning as an independent and self-contained video transmitting station.

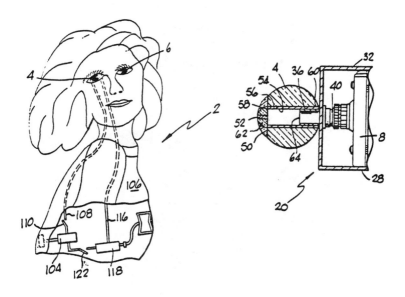

Lock Combination Decoder

Patent Number: 4,905,490
Date of Patent: Mar. 6, 1990
Inventor: Glenn E. Wilson, Endicott, NY

This device provides a nondestructive, automatic, and rapid technique for opening a safe whose combination is unknown. The decoder mechanism is built on a frame that can be attached to the safe with straps or bands. A computer-controlled DC servo motor (*44*) drives a cylindrical gripper (*40*) that fits over the knob of the dial. The dial position at any time is sensed by a high resolution rotary encoder (*48*), which can determine its orientation to within one-one thousandth of a revolution. A sensitive microphone (not shown) attached to the safe body picks up the sound of the faint click produced when the dial is rotated to a contact point. The computer is typically able to find the contact points and decipher the combination of a three-digit safe in about forty minutes.

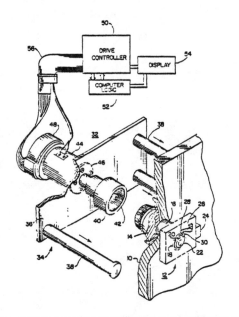

Motorcycle Safety Apparel

Patent Number: 4,825,469
Date of Patent: May 2, 1989
Inventor: Dan Kincheloe, San Clemente, CA

The survivability and degree of serious injury suffered in a motorcycle accident depend strongly on the type and amount of protective clothing worn by the rider. Boots and heavy leather clothes may be quite protective against abrasion, but provide relatively little impact protection. More effective protective suits have been devised, but because of their spacesuit-like appearance and confining characteristics, they are not in widespread use. This safety apparel may be made in stylish and unencumbering designs to encourage regular use by motorcycle riders. In the event of an accident, a source of compressed or liquefied gas rapidly inflates the suit to form a protective enclosure for the parts of the body most susceptible to critical injury. When manufactured as a jacket, a hood expands from under the collar to surround the head while a lower section expands downward to below the knees.

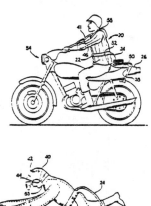

Theft Protection Purse

Patent Number: 4,804,122
Date of Patent: Feb. 14, 1989
Inventor: Renior L. Knox, Corpus Christi, TX

At first glance, this theft protection purse appears to be a conventional-looking purse (*11*) with a carrying handle. A flexible flap (*13*) secured by Velcro conceals a secondary purse (*19*), which is attached to the user's wrist by a bracelet and tether line. All of the user's money and valuables are stored in the secondary purse, while the main purse is used to transport everyday necessities. Should the main purse be stolen, the secondary purse will be ripped from its concealed location by the tether line and remain in the possession of the user.

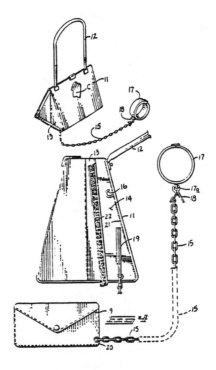

Remotely Controlled Briefcase Alarm

Patent Number: 4,804,943
Date of Patent: Feb. 14, 1989
Inventor: Isaac Soleimani, Jamaica, NY

In the event that this briefcase is taken from the possession of an authorized carrier, the owner can activate a radio-controlled alarm system within the briefcase. The alarm can be a siren or harmless smoke. At the same time, an electric solenoid can be activated, causing the handle to separate from the briefcase. This would result in the briefcase falling to the ground, leaving the thief with only the handle. In another variation of the invention, the handle could be made to painfully clamp onto the hand of the thief, providing a strong inducement to drop the briefcase.

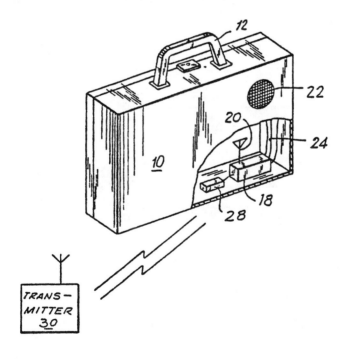

Beach Safety Anchor Security System

Patent Number: 5,199,361
Date of Patent: Apr. 6, 1993
Inventor: Milton W. Robinson, Media, PA

This beach safe provides a theft-proof container to secure valuables while swimming, napping, or taking a walk. It could also be used by fishermen, campers, and outdoor workers. To install the safe, an anchor shaft (*11*) having an auger tip (*20*) is embedded into the ground by rotating a handle (*14*) at the top. The handle is then removed, an anchor pan (*10*) is slipped over the end of the shaft, and the handle is replaced. The anchor pan may be filled with sand for extra weight and security. A lockable storage container (*6*) is set into the pan and held in place by pins (*28*) with quick-release clips (*29*). Appropriate hardware may also be added to support a beach umbrella.

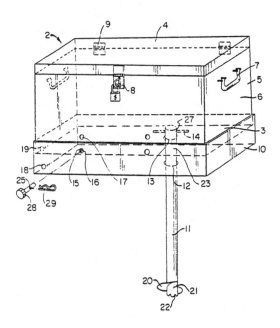

Protective System for Protecting Against Assaults and/or Intrusions

Patent Number: 4,821,017
Date of Patent: Apr. 11, 1989
Inventors: Yair Tanami
          Yoav Madar
          both of Gedera, Israel

This system is intended to protect a motor vehicle driver from assault or threat of violence by a passenger. A pair of electrodes (2, 4) are placed over the front and rear passenger-carrying seats. Each electrode is connected to a high voltage ignition coil (6, 8) that may be energized by a battery (10) and a set of motor-driven breaker points (14). Selector switches (18, 20) allow the driver to place either the front, rear, or both high voltage coils on line. If the driver is threatened, operating a foot switch (22) subjects the passenger to a very severe shock of up to 60,000 volts, sufficient to immobilize him and to cause him to drop his weapon if he is holding one. The system may also be used in an automatic mode to protect the vehicle from theft when it is left unattended.

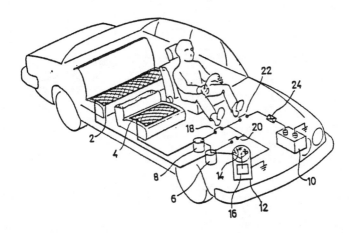

Ballistic Stream Electrical Stunning Systems

Patent Number: 4,930,392
Date of Patent: June 5, 1990
Inventor: John R. Wilson, Patterson, CA

This security system is intended to be used in antitheft, antiintrusion, antiskyjacking, and riot control applications. An electrically conductive liquid is stored in a pair of reservoirs. The liquid may be ordinary tap water containing salt ions and a viscosity increasing agent that allows a stream to hold together over long distances without breaking up into spray. A number of sensors monitor the area to be protected. When unauthorized entry is detected, two parallel streams of the conductive liquid (22) are projected at the intruder from insulated manifolds (15). A high voltage power supply capable of delivering 40,000 volts is applied across the manifolds, stunning and immobilizing the intruder.

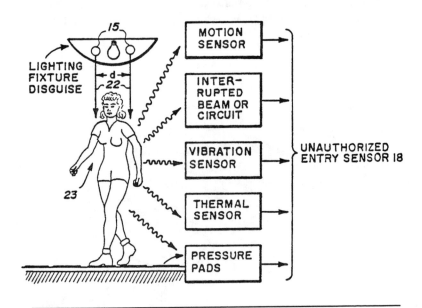

Attack Prevention Method

Patent Number: 5,137,176
Date of Patent: Aug. 11, 1992
Inventors: Paul P. Martineau
          Anne E. Martineau
          both of Pocasset, MA

This attack protection device consists of a wax capsule that is carried in the user's mouth. The capsule contains a hollow cavity filled with citric acid. In the event of a threatened attack, the user rapidly bites down on the capsule, releasing the acid solution into the mouth. The user then expectorates the acid into the eyes and face of the attacker. The acid solution is claimed to be nonharmful to the internal skin of the user's mouth.

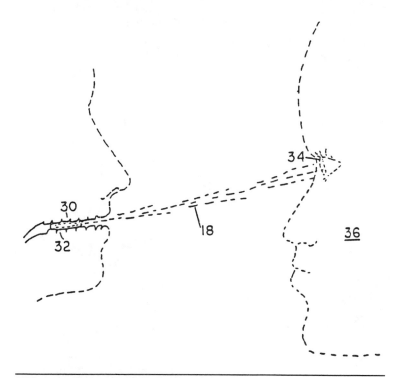

Jogger's Nightstick

Patent Number: 4,957,057
Date of Patent: Sep. 18, 1990
Inventor: Albert Marcucci, Mississauga, Ontario, Canada

The sight of a jogger moving quickly often incites animals, particularly dogs, to give chase and sometimes attack. This defensive baton can be carried without discomfort or inconvenience by a jogger. It is intended to protect the user from animal attack without unduly harming the animal. In the event of an attack, the user grasps the molded handle (*16*). The striking portion of the baton (*14*) is soft and resilient, so that should it be necessary to strike an animal, no permanent damage is likely to result. It is also attached by means of a break-away joint that will give way should excessive striking force be used. A rattle (*40*) is built into the baton to confuse and scare off animals. Reflectors (*26, 28*) also make the jogger more visible to traffic at night. No mention is made regarding use of the nightstick for protection against human assailants.

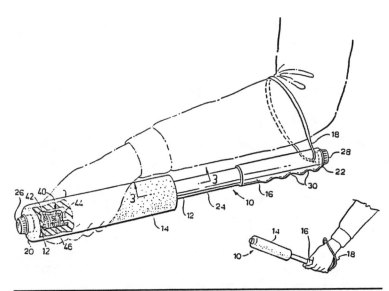

Dog-Tracking Scent-Dispensing System for Apprehending Burglars and the Like

Patent Number: 4,867,076
Date of Patent: Sep. 19, 1989
Inventor: **Louis J. Marcone**, Rochester, NY

This security system is used to spray a person engaged in a bank hold-up with a chemical scent that is readily tracked by a trained police dog. The scent is a clear, non-toxic liquid spray that is odorless to humans. One satisfactory scent formula contains shark liver oil, vegetable oil, pylam LX5880, pylakrom oil, and butyric acid. In the example shown, a bank teller releases the spray from a pressurized aerosol can by depressing a foot switch (*38*). With the addition of suitable intrusion sensors, the system can also be adapted for automatic protection of windows, doors, passageways, and fire alarm call boxes.

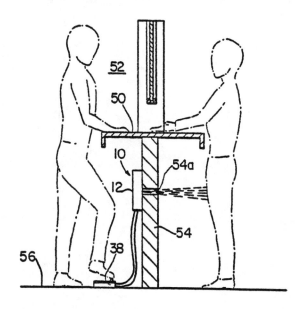

Neonatal Net

Patent Number: 4,963,138
Date of Patent: Oct. 16, 1990
Inventors: Nohl A. Braun, Jr.
Mary J. Urling
both of Charleston, WV

Safe handling of a newborn baby immediately after delivery is not always easy because the infant is covered with a slippery fluid. This invention is intended to enhance the safety of the newborn child while it is being handled by the obstetrician or the attending nurses. The net may be used in several ways. In one form of the invention, it is connected between the end of the delivery table and the waist of the obstetrician's surgical gown by means of Velcro straps. In another form, it spans the gap between the delivery table and the neonatal examination table.

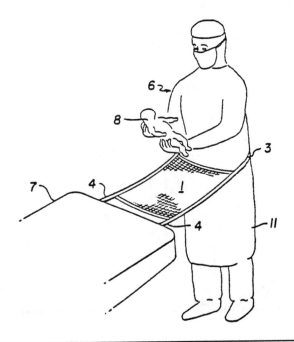

Newborn Antitheft Device

Patent Number: 4,899,134
Date of Patent: Feb. 6, 1990
Inventor: Clifford R. Wheeless, Jr., Baltimore, MD

This umbilical cord clamp contains a device that allows the detection of unauthorized movement and possible abduction of a newborn child. It operates on the same principle as some antitheft tags used to safeguard merchandise in retail stores. An AC magnetic field is emitted from a transmitting loop in the doorway of the maternity ward. This magnetic field interacts with a wire loop (*20*) built into the umbilical cord clamp. A receiving loop in the doorway detects the modified magnetic field when the clamp (and child) is moved close to the doorway. An alarm is then activated.

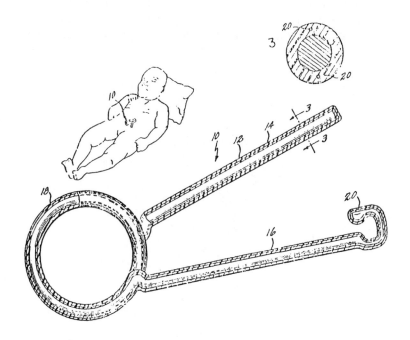

Ink- or Dye-Filled Blister Packs

Patent Number: 5,208,085
Date of Patent: May 4, 1993
Inventor: Marvin B. Pace, Ludlow Falls, OH

A common problem faced by mailbox owners is that vandals will often drive by and strike the mailbox with a baseball bat. Many homeowners who live in rural areas typically experience such vandalism several times a year and the cost and time involved in replacing mailboxes quickly becomes intolerable. This invention provides a means for deterring vandalism to mailboxes and other structures. It consists of a self-adhesive plastic panel covered by a large number of dye-filled nodules (*16*). Each nodule is about 1 inch long and has a maximum height of about ¼ inch. When a nodule is struck, the combination of its triangular profile and a pre-weakened burst seam causes it to eject its contents toward the vandal. The preferred dye composition is a mixture of a permanent dye and petroleum jelly, which allows the dye to travel forward as a coherent mass rather than dispersing laterally.

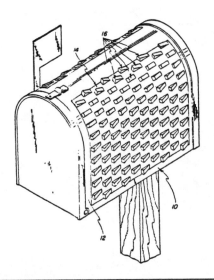

Combination Hand and Finger Cuff

Patent Number: 5,230,351
Date of Patent: July 27, 1993
Inventor: Sahr A. A. Nyorkor, Indianapolis, IN

This improved restraint device is intended to restrict the movement
of a person's hands. It will prevent them from holding an object such
as a gun or a knife, which might be used to cause injury to
themselves or others. It consists of a pair of rigid metal or plastic
gloves that are secured to a person's wrists with locking handcuffs.
A short metal chain keeps the person's hands close together. No
mention is made regarding the possible injury that could be inflicted
using the rigid gloves themselves as weapon.

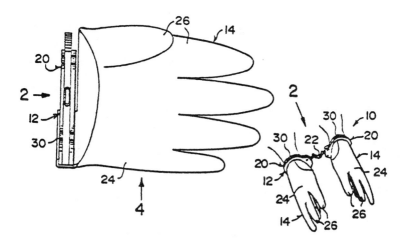

Sports and Exercise

Something about our competitive nature compels us to turn almost everything we do into a sport or contest, and this provides fertile ground for new inventions. Golf accessories, in particular, receive a large number of new patents each month. Many of these patents are concerned with training devices to improve one's golf swing, but one enterprising golfer even patented his own golf swing itself! Fishing lures generally run a close second to golf clubs. One relatively simple casting lure patent had forty-two claims regarding its uniqueness! In this section we also find exercise machines devised to strengthen every part of the human body.

Thumb-Wrestling Game Apparatus With Stabilizing Handle

Patent Number: 4,998,724
Date of Patent: Mar. 12, 1991
Inventor: Richard B. Hartman, Issaquah, WA

This improved apparatus for playing a game of thumb wrestling includes a stabilizing handle. The handle employs forces generated by firmly interlocking the fingers of the players' hands to stabilize and anchor the miniature game ring. The game ring surface includes holes through which the thumbs of the opponents are inserted and a solid wrestling region upon which one thumb can forcibly pin another thumb, thereby giving the realistic impression of a pin in wrestling.

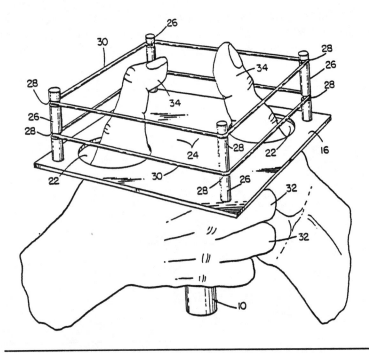

Toe Exercise Device

Patent Number: 4,869,499
Date of Patent: Sep. 26, 1989
Inventor: Donald R. Schiraldo, Bayonne, NJ

This dynamic splint is designed to stretch the tendons of the big toe as well as to exercise the nearby muscles and tissues. A relatively stiff semicircular brace (*12*) is held above and behind the ankle by a Velcro strap (*14*). Elastic rubber tubing (*20*) is connected to the brace, as well as to a toe pouch (*22*). In a manner similar to drawing a slingshot, the big toe is thrust forward to stretch the rubber tubing. According to the inventor, use of the device will speed recovery from toe injuries.

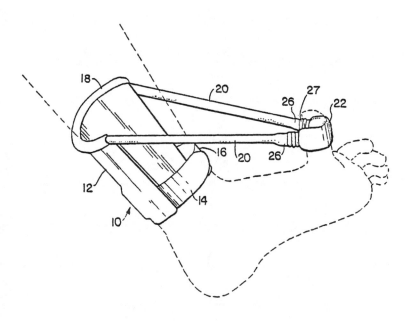

Neck-Exercising Device and Method

Patent Number: 4,832,333
Date of Patent: May 23, 1989
Inventor: Ricky P. Lockett, Philadelphia, PA

This invention allows all of the muscles of the neck to be exercised without removing the device. A cushioned headband is secured tightly, but comfortably, around the head with a Velcro fastener. The headband has a number of loops around its circumference. To exercise a particular group of neck muscles, the user attaches an elastic cord by means of a snap hook to one of the loops. The user may then pull on the cord while attempting to hold the head in place, or may move the head while holding the hand in place. The snap hook is easily moved to a different location to exercise a new group of muscles.

Abdominal Exerciser

Patent Number: 4,775,148
Date of Patent: Oct. 4, 1988
Inventor: Gary G. McLaughlin, Arleta, CA

This abdominal exerciser is attached to a reclining user's waist by means of a belt. A contacting plate (*18*) and a retainer plate (*20*) are held apart by a spring (*22*). Exercise is achieved by alternately tightening and releasing the muscles of the abdomen. This presses and releases the contacting plate, causing the abdominal muscles to work against the spring. If desired, an additional weight can be added to the underside of the contacting plate, which compresses the abdomen inwards, providing additional travel and resistance during the exercise.

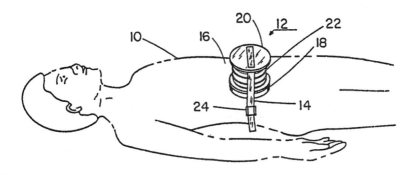

Body-Supported Resilient Exercise Apparatus

Patent Number: 4,911,439
Date of Patent: Mar. 27, 1990
Inventor: Larry L. Kuhl, Mason City, IA

This exercise harness is made from a length of elastic cord formed into two loops. After putting on the harness, a user grasps a pair of handles and performs exercises by extending the arms and stretching the elastic cord. Additional handles can be provided to allow a wide variety of exercises to be carried out. Since the harness is extremely lightweight, it could also be used while running, jogging, or walking.

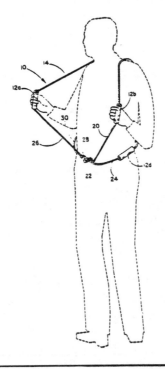

Throwing Arm Exercising Machine

Patent Number: 5,197,933
Date of Patent: Mar. 30, 1993
Inventor: Tommy R. Waters, Meridian, MS

This improved exercising machine not only increases the strength of the user's throwing arm, but also provides guidance as to the proper form for throwing. It can be used equally well by right- or left-handed throwers. To use the machine, the user grasps a T-shaped handle that slides in an arced overhead track. The curve of the track is tailored to closely conform to the correct throwing motion for the sport being practiced (baseball, football, etc.). An adjustable resistance of between 3 and 20 pounds is provided by a weight (*34*) and pulley (*37*) system. To increase user stability during exercising, a grip is provided on the frame for the nonthrowing hand.

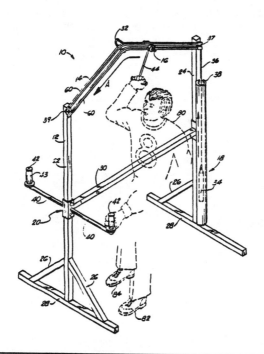

Engine-Spiraled Stabilized Toy Football

Patent Number: 4,923,196
Date of Patent: May 8, 1990
Inventor: Daniel Z. Rohring, Norman, OK

This toy football contains a propulsion system for increasing the length of its trajectory while simultaneously providing gyroscopic spin stabilization of the ball during its flight. Inside a football-shaped plastic skeleton, a miniature internal combustion engine (*70*) is used to power a propeller (*82*) with a gyroscopic ring (*92*) around its circumference. The skeleton is covered by a plastic skin, except at the two ends to allow for air intake (*20*) and exhaust (*22*). Fuel is contained within a small-diameter flexible bladder (*108*). To start the engine, a 1.5 volt battery is connected to the ignition wire (*112*) to energize a glow plug (*74*). A cordless drill with an elongated starting chuck (not shown) engages the propeller nut (*86*) and rotates the engine until it fires. The engine is connected to the skeleton by its mountings so that as the propeller turns, the spinning motion provides a gyroscopic effect that stabilizes the ball during its flight.

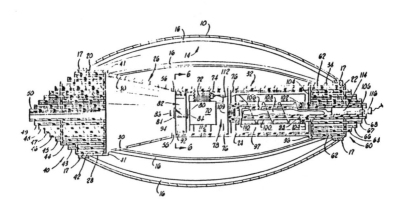

Ice-Skating Exercise Device

Patent Number: 4,781,372
Date of Patent: Nov. 1, 1988
Inventor: Patrick J. McCormack, Hull, MA

This exercise apparatus allows a user to practice a wide variety of actual ice skating and ice hockey movements. The user's feet are strapped into stirrups that are free to slide and rotate on a pair of grooved tracks. The space between the tracks can be set at any angle from 0 and 180 degrees in order to simulate a number of different skating positions. Resistance to leg motion is provided by a weight and pulley system. Resistance to ankle motion is provided by a friction device within each stirrup pivot.

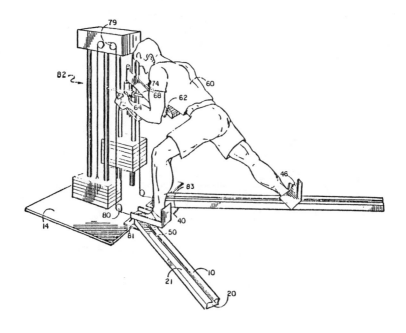

Parachute for Skaters and Runners

Patent Number: 5,217,186
Date of Patent: June 8, 1993
Inventors: Lloyd G. Stewart, Memphis, TN
Kirk A. Stewart, Germantown, TN
Dean Lotz, Bartlett, TN

Runners frequently carry weights to increase their strength and endurance. The distribution or attachment of weights to the body presents problems and they are heavy and inconvenient to transport. This invention provides a lightweight and easily portable means for imparting a drag to a runner, speed skater, bicyclist, or other rapidly moving person. It consists of a parachute clipped to a belt worn by the user. The area of the parachute will vary depending on the size, weight, and speed of the user. A hole in the center of the parachute stabilizes it and prevents it from swaying side to side. According to the inventors, the device may also be used on animals such as racing dogs and horses.

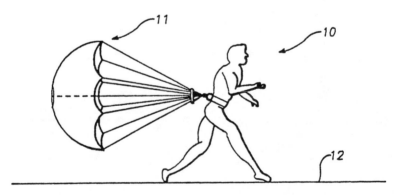

Jogger's Aid

Patent Number: 5,207,719
Date of Patent: May 4, 1993
Inventor: Ronald E. Janus, Deerfield Beach, FL

One of the major problems encountered in many exercise activities is the inability to conveniently carry a supply of liquid to replenish lost bodily fluids. Previous devices have stored fluid in a container located too far away from a user's mouth, requiring a great deal of suction to ingest the liquid. This insulated fluid container is secured around a user's neck to provide a readily accessible supply during exercise. Adhesive on the back side of the container sticks to the user's clothing to reduce movement during vigorous activity such as running or bicycling. Liquid is sipped through a flexible straw.

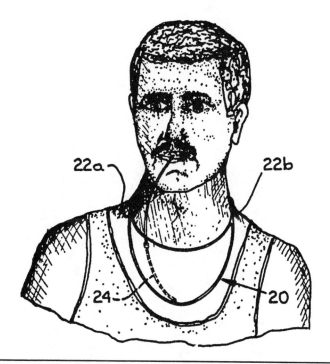

Guidance Apparatus for Bicycle Training

Patent Number: 5,154,096
Date of Patent: Oct. 13, 1992
Inventors: Rami Geller, Berkeley, CA
　　　　　　Avry Dotan, El Cerrito, CA

This guidance apparatus simplifies the age-old problem of teaching a child to ride a two-wheeled bicycle. The elevated handle allows a supervising adult to easily control the balance of an inexperienced bicycle rider. In case of an emergency, the adult can slow or stop the bike quickly by using an additional handbrake. The apparatus is clamped to the frame of a standard bicycle with a pair of U-bolts *(20)*.

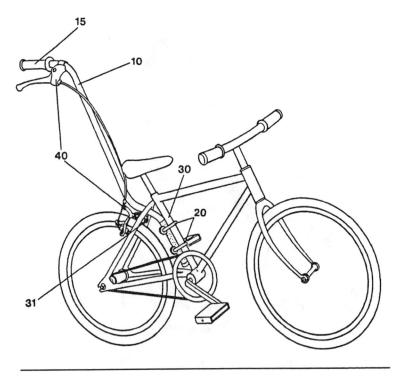

Body-Supported Hoop Game and Device

Patent Number: 4,871,178
Date of Patent: Oct. 3, 1989
Inventor: Wilfredo S. Diaz, Guayama, Puerto Rico

In this game, the goalies of two opposing teams each wear a body-supported hoop. The hoop is attached to the goalie's knees by a pair of leather or elastic girdles. The object of the game is for team members to place a projectile (such as a soccer ball) through their own goalie's hoop and to prevent the opposing team from scoring in a similar manner. The inventor suggests that a conventional crotch guard (56) preferably should be worn to protect the goalie's groin area from impact with the ball. Other guards for the chest and face may also be necessary.

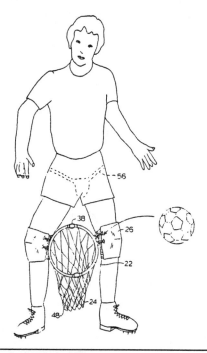

Training Apparatus for Cattle Roping

Patent Number: 5,192,210
Date of Patent: Mar. 9, 1993
Inventors: Darrell E. Thomas
Donald E. Standley
Leon J. Jackson
all of Anahuac, TX

In competitive cattle roping, many hours are spent practicing roping and forming a "dally," which involves wrapping an end of the rope around the saddle horn without entangling one's fingers. The dally enables the rider to control a roped animal and also to release it quickly if necessary. In this improved training apparatus for cattle roping, the user wears a frame assembly strapped about the hips. The frame supports a saddle horn mounted in the same relative position that it would occupy on an actual saddle. The device permits a user to practice roping a steer dummy from a variety of positions and approaches. It also permits the user to safely practice forming a dally until the maneuver becomes second nature.

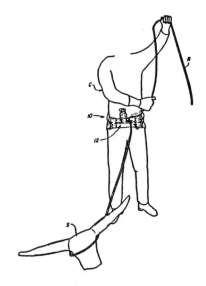

Training Horse Simulator

Patent Number: 4,957,444
Date of Patent: Sep. 18, 1990
Inventor: Seth A. Armen, Weston, CT

This invention provides a means to practice basic horseback-riding skills such as mounting and rein handling without an actual horse. The horse's body is constructed of a pivoting head assembly and a barrel fitted with a contoured seat or saddle. An easy-to-grip handle (*34*) replaces the usual pommel. A pressure-sensitive switch in the saddle activates an indicator light to inform the rider when sufficient contact has been made with the saddle during posting. Other switches monitor the level of tension applied to the reins and activate indicators corresponding to rein commands for left turn, right turn, and stop. It permits the user to discover when conflicting or confusing commands are being given to the simulated horse. The inventor suggests that the device would be especially useful for teaching children or handicapped people to ride.

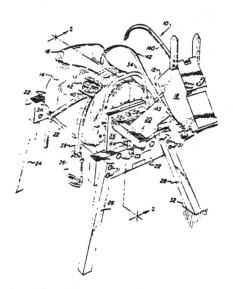

Electronic Martial Arts Training Device

Patent Number: 4,974,833
Date of Patent: Dec. 4, 1990
Inventors: Kenneth D. Hartman, DeKalb, IL
Steven A. Overmyer, Elgin, IL

This martial arts training device is constructed to be rigid enough to simulate the feel of a human body when struck, but yet pliable enough to ensure that the user will not be injured when delivering a full impact blow. Its surface contains a pictorial representation of a martial arts combatant. A target light is mounted underneath a layer of flexible transparent potting material at each of the pictured combatant's "vital points." At the location of each light, an embedded miniature transducer generates an electrical signal when struck. For a particular skill level, a martial arts student is given a certain amount of time to strike a vital point target light after it is illuminated by a control circuit. As the skill level is increased, the time is shortened. Hits are recorded on a digital counter.

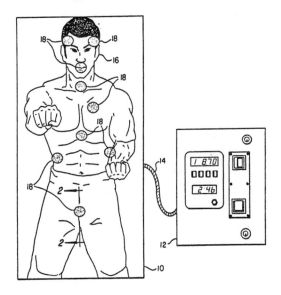

Wing Apparatus for Skiers

Patent Number: 4,890,861
Date of Patent: Jan. 2, 1990
Inventor: William V. Bachmann, St. Clair Shores, MI

This apparatus provides aerodynamic lift and a sensation of flying to a downhill skier. Two independent airplanelike wings are attached to a harness worn by the user. Each wing can be independently rotated by means of a handgrip (*124, 124a*). A spring-loaded torsion bar returns the wings to a preset angle when pressure is removed from the controls. Varying the angle of attack of the wings allows the user to initiate turns or slow down. The lightweight fabric of the wing panels can also be collapsed about the central spars, allowing them to be used as ski poles.

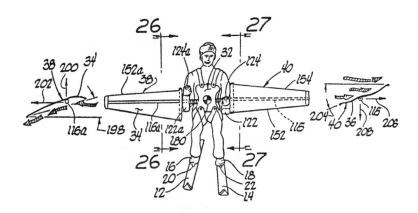

Indoor Ski Slope and Apparatus for Making Snow Thereon

Patent Number: 4,790,531
Date of Patent: Dec. 13, 1988
Inventors: Nobuyuki Matsui   Tadashi Yoshida
         Shinichi Yokota   Hachiro Nonaka
         Kazuo Otsuka   Tsutomo Okumura
         Shuhei Mizote   all of Tokyo, Japan

Previous artificial ski slopes have been limited by the amount of land available, resulting in very short and unchallenging runs. Furthermore, large energy costs are incurred in temperate climates when cooling the entire containment building of an indoor slope. This improved indoor ski slope solves both of these problems by constructing the run underground, supported by one or more massive columns. The structure is built in a helical or figure eight pattern, which provides the necessary elevation change, as well as a long continuous path from top to bottom. By leaving only the ski lodge above ground, the insulating properties of the earth can be used to substantially reduce energy costs. The snow-making machinery is contained in a track-mounted module that travels along the surface of the run. By incorporating inflatable, frozen, water-filled mats in the surface of the run, the capability is provided to simulate a wide variety of conditions encountered in natural outdoor skiing, such as random moguls or an entire mogul field.

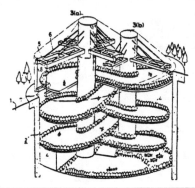

Pole Vault Simulator Device

Patent Number: 4,778,174
Date of Patent: Oct. 18, 1988
Inventor: Brant Tolsma, Forrest, VA

The pole vault exercise is one of the most difficult events to master in track and field. Its elements consist of a run, takeoff, rock-back, and clearance of the bar. This simulator allows an athlete to practice tilting his or her body back and clearing the bar without the necessity of performing the energy-draining run of the approach phase. It permits a coach to provide continuous guidance to an athlete trying to develop improved "in-flight" technique. The vaulter grasps the short bar (26), jumps upward, and realigns into an upside-down position. As he or she begins to drop downward, his body weight applies a load to a spring assembly (6). An assistant pulling on a handle (34) supplies an additional upward force that works in conjunction with the spring assembly to propel the vaulter into the air. Another spring (30) and cable (28) simulate the mechanics of the pole. Upon reaching the desired height, the vaulter reorients and pushes his or her body over the clearance bar (36).

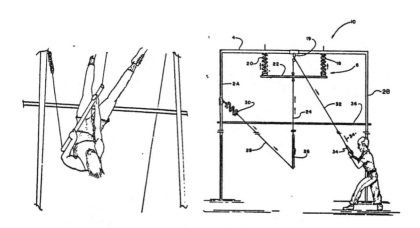

Equipment for Towless Skiing on Water Surface

Patent Number: 4,804,345
Date of Patent: Feb. 14, 1989
Inventor: Jong S. Lee, Daejeon, Choong-nam,
Republic of Korea

This invention comprises a pair of propulsion sticks and floating foam-filled water skis. The user's foot is inserted into a shoe (*80*) at the bottom of a cavity in the ski. The bottom of each ski is shaped to facilitate walking into the water from the shore as well as for surf skiing on shallow, rough, or deep water. The two skis are connected by lightweight ropes attached to a pair of eyelets (*70, 72*) at the front and rear. This adds stability and helps to prevent the skis from drifting apart laterally. The propulsion sticks are tipped with buoyant structures designed to push against the water and be withdrawn with a minimum of drag. The sticks can also be mounted crosswise across the skis to form a rigid catamaran structure. This facilitates boarding the skis from the water and provides a stable resting platform if the user becomes tired.

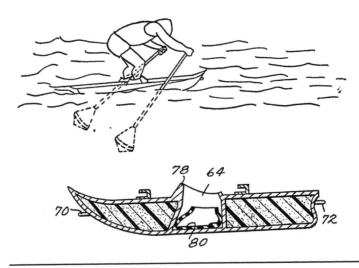

Dry Land Swimming Training Apparatus

Patent Number: 4,830,363
Date of Patent: May 16, 1989
Inventor: Robert J. Kennedy, Ottawa, Ontario, Canada

This swimming simulator permits a user to experience the motions and forces encountered during many different swimming strokes. It permits training for both strength and endurance in the front crawl, breast stroke, back crawl, and the butterfly stroke. The user's arms are exercised by a retractable pulley system, while the legs work against hydraulic shock absorbers (40).

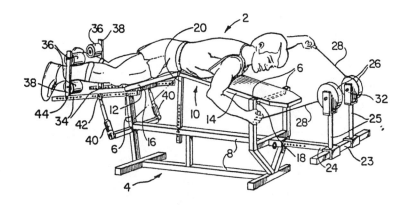

Swimming Aid

Patent Number: 4,832,631
Date of Patent: May 23, 1989
Inventor: Marvin N. Gag, North Mankato, MN

This swimming aid will support a beginning swimmer in the water and operates with arm motions that mimic a normal swimming stroke. Using an overhand motion, the user rotates a pair of handles, thereby providing power to a crankshaft (*16*). A pair of foamed-plastic paddles are connected to the ends of the crankshaft. As one paddle engages the water to provide propulsion, the other is disengaged and provides flotation only. There is no orientation restriction in that the device can merely be thrown into the water and used by the swimmer from either side.

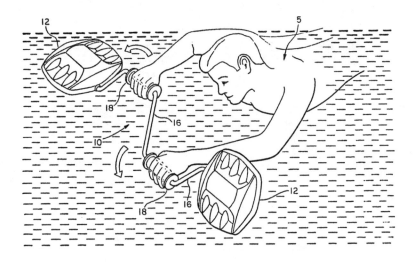

Personal Flotation System

Patent Number: 4,861,301
Date of Patent: Aug. 29, 1989
Inventors: Jimmy R. Pomeroy
　　　　　　Ruth N. Warwick
　　　　　　both of New Braunfels, TX

This personal flotation system allows for convenient storage of refreshments while floating on lakes, rivers, and mild surf. It consists of an inflated inner tube that supports the user. A cloth haversack is attached to the outside of the tube with a Velcro fastener. The haversack contains an ice chest and a zippered pocket (*311*) to carry small valuables such as keys or coins. A second Velcro fastener on a strap (*308*) prevents the ice chest lid from coming loose in rough seas.

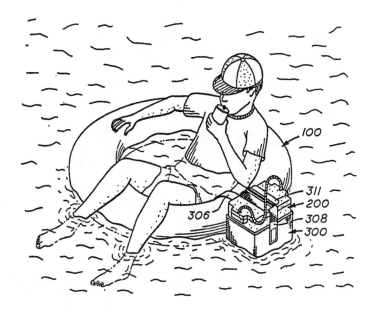

Aquacycle

Patent Number: 4,789,365
Date of Patent: Dec. 6, 1988
Inventor: Jeffrey K. Jones, Portland, OR

With this invention, a conventional bicycle can be quickly and easily transformed into an aquacycle. No modification of the bicycle is required. The conversion apparatus is constructed mainly from lightweight PVC tubing and connects to the bicycle with two U-bolts. A twin hull assembly supports the bicycle by the horizontal portion of its frame. The rear wheel of the bicycle rests on a friction wheel (68) which rotates twin propellers via dual flexible axles (74). The front wheel of the bicycle rests in a pivoting carriage (62). Dual rudders (42) are controlled by a cable (58) and pulley (50) system that turns the aquacycle in the same direction as the handlebars.

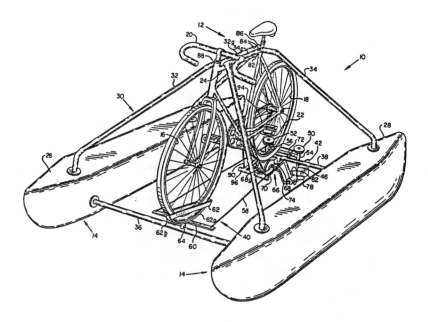

Remote Control Angling Devices

Patent Number: 5,154,016
Date of Patent: Oct. 13, 1992
Inventors: Gary W. Fedora, Georgetown, Canada
Douglas Sehl, Crystal Beach, FL

For the shore-bound angler, the ability to accurately position a lure or bait is a function of casting skill, the ballistic properties of the fishing tackle, and the strength of the fisherman's arm. Taken together, these factors constrain an angler's fishing activity to a relatively small area of water in the angler's immediate vicinity. This remotely controlled fishing boat is operated by a hand-held radio transmitter (*3*) and enables an angler to fish a large area of water. The terminal tackle (*9*) can be positioned at a desired depth by means of a motorized downrigger mechanism (*10*). The fishing line is released from the downrigger by a sharp tug, either from the angler or a fish. A downward-looking sonar (*7*) reports bottom depth and fish activity to the user by means of an array of lights on the mast (*8*).

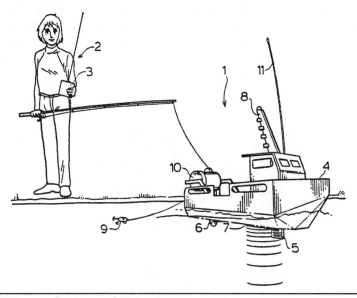

Electrofishing Pole

Patent Number: 5,214,873
Date of Patent: June 1, 1993
Inventor: Norman G. Sharber, Flagstaff, AZ

The user of this electrofishing pole holds an insulated fiberglass wand that terminates in a submerged stainless steel loop. The loop acts as the anode (+ electrode) of an electrical circuit and is protected from direct contact with the fisherman or the fish by a number of water-permeable "whiffle" golf balls strung around its circumference. A backpack supports a power source (*20*) and a pulse-generating circuit (*22*). A cathode (− electrode) wire with an uninsulated end (*26*) trails in the water behind the user, completing the electrical circuit. The voltage between the anode and the cathode may be in the range of 500 to 1000 volts. Any fish within a sufficiently strong portion of the electric field created by the electrodes will experience involuntary muscle responses that actually cause it to swim closer to the loop. At some point it will lose consciousness and can be netted. The inventor suggests that the fisherman might also consider wearing rubber gloves. It should also be noted that this method of fish harvesting is used to capture endangered species for tagging and release.

Method and Apparatus for Temporarily Immobilizing an Earthworm

Patent Number: 4,800,666
Date of Patent: Jan. 31, 1989
Inventor: Loren Lukehart, Boise, ID

The ability of an earthworm to curl its body in almost any direction, coupled with the fact that it is coated with a slimy film, makes it extremely difficult for a fisherman to impale the earthworm with a fishing hook. This invention provides a means for "dewiggling" an earthworm to facilitate the baiting process. To dewiggle a worm, the user places it into a rectangular container that is partially filled with sharp sand having a grain size of less than or equal to one-twentieth of an inch. The sand becomes embedded in the earthworm, causing the worm to immediately relax. After the worm is baited on the hook and immersed in the water, the sand rinses free and the worm's normal wiggly character is restored.

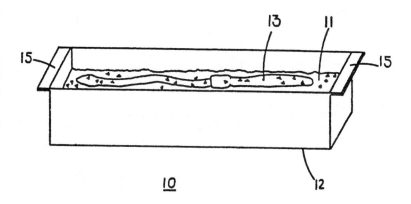

Fisherman's Float-Propelling System

Patent Number: 4,938,722
Date of Patent: July 3, 1990
Inventor: Harold K. Rizley, Sayre, OK

Fishing floats are widely used and are generally propelled by swim fins attached to the fisherman's feet. To avoid overtiring the fisherman and to increase the enjoyment of float fishing, this invention provides an economical powered form of propulsion. With the fisherman seated in the float, a commercially available electric trolling motor (*14*) is strapped to the calf of one leg. The motor is powered by a storage battery located in a Styrofoam container (*30*) that rides in a small auxiliary float. When the fisherman desires to move to another location, he extends his leg horizontally in the direction that he wishes to go and activates the motor switch (*36*). The fin on the fisherman's other foot acts like a rudder and provides additional steerage.

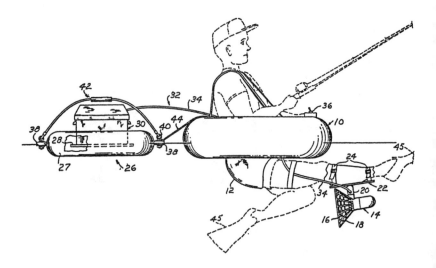

Fishing Lure Apparatus

Patent Number: 5,133,148
Date of Patent: July 28, 1992
Inventor: Michael J. Lawson, Ayer, MA

This radio-controlled fishing lure is intended to simulate a hovering dragonfly to attract bass and other gamefish. It is powered by a small motor (*13*) within the body that turns a propeller. A radio receiver (*18*) operated by battery (*17*) controls servo mechanisms (*19, 20, 21*) that adjust the throttle speed, tail elevator (*15a*), and rudder (*16a*).

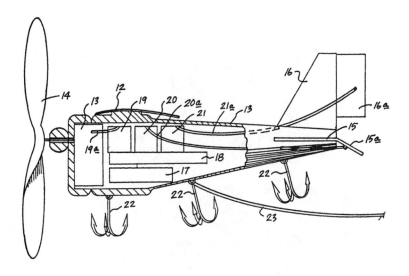

Tracer Golf Ball

Patent Number: 4,564,199
Date of Patent: Jan. 14, 1986
Inventor: James S. Adams, Houston, TX

This tracer golf ball enables its user to easily observe the characteristics of the ball in flight. It has a center chamber filled with a smoke-producing agent. The body includes a release valve for chemical smoke, which trails the ball in flight after it has been struck. The core of the ball also includes air chambers that enable the ball to float in water.

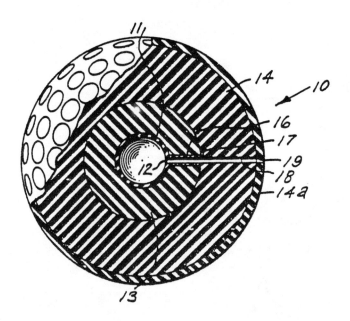

Golf Swing Training Apparatus

Patent Number: 5,050,885
Date of Patent: Sep. 24, 1991
Inventors: James T. Ballard, Miami, FL
             Norlin O. Lewis, Remlap, AL
             Carlton W. Montgomery, Clay, AL
             Charles H. Birdsong, Meridian, MS

Over the years, a number of patents have been issued for apparatus to train golfers in the "perfect" swing. This trainer induces a golfer to make the "Ballard Swing," which features a combination of lateral and rotational hip motion. The golfer wears a hip saddle that limits and guides hip movement. Various springs and elastic devices urge proper coordination of the hip saddle and a rigid shoulder plate attached to a vest. An adjustable leg stop acts as a target for the right leg at the reversal point from backswing to downswing and limits the lateral motion and degree of weight shift. The golfer can actually strike the ball while being fully guided by the apparatus in order to learn proper technique.

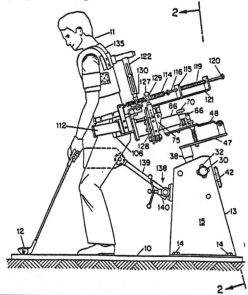

Golf Training Putter

Patent Number: 5,213,331
Date of Patent: May 25, 1993
Inventor: Frank Avanzini, Winter Park, FL

Accurate putting demands precise alignment between the putter and the cup. This putter includes a laser unit (*30*) mounted on the putter head to indicate the direction that the ball will travel when struck. The laser is activated by a button (*36*) in the handle of the club. The golfer can determine the initial alignment of the club by observing the point (*56*) at which the laser beam (*38*) strikes the ground. As the club is moved during the putting stroke, the laser beam traces out a line (*40*) to the cup. The device may also be provided as a retrofit kit for installation on the user's favorite putter.

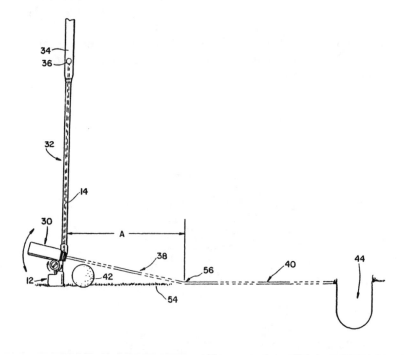

Putter

Patent Number: 5,131,660
Date of Patent: July 21, 1992
Inventor: Joseph Marocco, Hawthorne Woods, IL

One of the most difficult aspects of putting is learning to develop an accurate feel for how far the ball will go when struck. This improved training putter provides a mechanism for measuring the backstroke of the club head from perpendicular and relating it to the expected distance of the putt. It uses a weighted encoder wheel (*30*) as a pendulum that rotates as the club head is drawn backward. An optical interrupter (*38*) senses the position of stripes (*40*) on the encoder wheel to determine its orientation. A microprocessor (*42*) calculates the club head speed based on the rate of rotation of the wheel. It also determines the estimated distance that the ball will travel based on user inputs regarding the slope and resistance of the green. The distance is read out on a liquid crystal display.

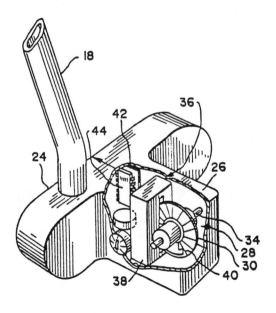

Skyboard

Patent Number: 4,898,345
Date of Patent: Feb. 6, 1990
Inventor: Dan Clayton, Rancho Cucamonga, CA

This skyboard is a combination of surfboard and parachute. It enables a user to ride the air currents in the sky in a much more exciting manner than conventional sailplanes and hang gliders allow. The skyboard has the general shape of a surfboard, but it is modified to include side wings for additional lift and fins on the underside for stability and steering. A pair of shoe bindings can be quickly detached by pulling a release cord (*60*). The skyboard has its own parachute, which can be triggered by the user at any time by pulling the second cord (*74*). If atmospheric turbulence causes the board to become unstable and the user is forced to eject, the board's parachute will bring it to a soft landing independent of the rider. According to the inventor, the user can stay in the air for an extended period of time and eventually glide to a safe landing on the ground or a body of water.

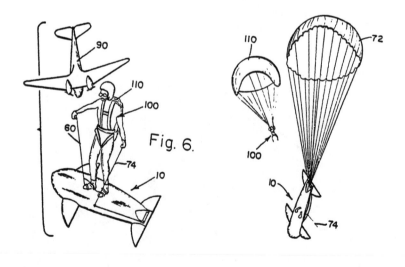

Fig. 6.

Cable Skydiving

Patent Number: 5,224,425
Date of Patent: July 6, 1993
Inventor: Bruce Remington, Eureka, CA

Cable skydiving provides a means to hurtle a rider downward to terminal velocity faster than any roller coaster. The ride can be experienced in any position, including standing, sitting, kneeling, prone, frontward, backward, upside-down, and even in the dark. The ride consists of two towers or other cable supports at different elevations with a cable loosely strung between them. A pulley block that is attached to some sort of passenger-carrying device rides on this cable. When the brake on the pulley block is released, the rider (*43*) travels downward at tremendous speed. After reaching the lowest point in the cable's arc, the rider continues uphill until stopping short of the end support by friction and gravitational means. A disembarking ramp (*42*) tilts upward to exit the ride. The inventor suggests that the ride may be made even more exciting by passing the cable through tunnels, buildings, waterfalls, fires, chasms, dark places, or other horrors.

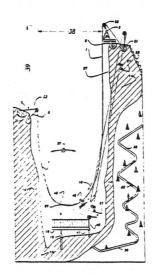

The Better Mousetrap

"Build a better mousetrap and the world will beat a path to your door."

—Ralph Waldo Emerson

Rodent Trap With Signal

Patent Number: 5,154,017
Date of Patent: Oct. 13, 1992
Inventor: Herbert R. Disalvo, Dunedin, FL

A unique feature of this spring-loaded rodent trap is its battery-operated signaling device. When the trap is sprung, an electrical circuit is completed, energizing the buzzer (27) and informing the user that a rodent has been dispatched.

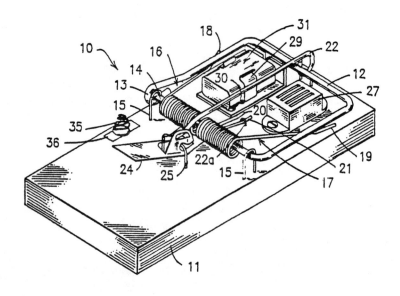